SpringerBriefs in Applied Sciences and Technology

Computational Mechanics

Magdalena Gromada · Gennady Mishuris
Andreas Öchsner

Correction Formulae for the Stress Distribution in Round Tensile Specimens at Neck Presence

 Springer

Magdalena Gromada
Ceramic Department CEREL
Institute of Power Engineering
1 Techniczna St.
36-040 Boguchwala
Poland
e-mail: gromada@cerel.pl

Andreas Öchsner
Department of Applied Mechanics
Faculty of Mechanical Engineering
Technical University of Malaysia
Johor
Malaysia
e-mail: andreas.oechsner@gmail.com

Gennady Mishuris
Institute of Mathematics and Physics
Aberystwyth University,
Penglais
Aberystwyth
SY23 3BZ Ceredigion
UK
e-mail: ggm@aber.ac.uk

ISSN 2191-5342
ISBN 978-3-642-22133-0
DOI 10.1007/978-3-642-22134-7
Springer Heidelberg Dordrecht London New York

e-ISSN 2191-5350
e-ISBN 978-3-642-22134-7

Cover design: eStudio Calamar, Berlin/Figueres

Printed on acid-free paper

Springer is part of Springer Science+Business Media (www.springer.com)

Preface

This monograph is devoted to one of the most popular methods for the determination of the plastic material properties, i.e. the tensile test particularly from the moment of neck appearance in the sample.

Despite the fact that a few different classical formulae exist which describe the stress distribution in the neck, there is not any certainty which of them is more accurate and the choice to apply a certain formula is often somewhat arbitrary. After thorough literature search, it turned out that the formula of Bridgman is more often utilised in practice for the yield stress determination. However, our preliminary analysis has shown that it may generate rather non-negligible error (up to ten percent) at least in the case of ideal plastic materials.

It is well known that in the western literature Bridgman's formula is more frequently used while the eastern literature prefers the approximation by Davidenkov-Spiridonova. Both of these formulae were derived in the forties of the last century. What is interesting is that for the first time a formula for the determination of the average normalised axial stress in the minimal section plane was derived by Siebel, which indeed overlaps with the approach proposed by Davidenkov-Spiridonova. Siebel's work is however less often used and its relative obscurity can probably be historically explained by the fact that it was published in Germany shortly after the Second World War. Obviously, repeated trials were made to derive more accurate formulae and at least two of them were successful (Szczepiński's, Malinin's) but the obtained solutions are still seldom utilised in practice also because of a lack of information on their accuracy in comparison with the classical formulae.

The authors' aim in the presented monograph is to collect all known results in the area and to answer the aforementioned questions. Indeed, one can find in the detailed description of materials flow curves determination, criteria of neck creation, derivations of all known formulae for stress distribution in the neck of axisymmetric samples as well as estimation of accuracy of simplifying assumptions applied during the derivation of the classical formulae by means of very accurate numerical simulations. As a result of the critical analysis of the simplifications, a new empirical formula was derived which depends on the same geometrical parameter (i.e. ratio of the sample radius in the minimal section to the contour radius of the deformed sample) as the classical formulae, but revealing higher accuracy than them. In addition, a new analytical model was proposed, which describes the stress distribution in the neck of an axisymmetric tensile specimen and on its basis a new formula for the average normalised axial stress in the minimum section plane was derived. This formula takes into account in addition to the mentioned parameters a new ratio (i.e. the relative neck radius in a measure of its deformation). Fortunately, both aforementioned parameters can be easily measured in experimental tests. During the verification of all formulae based on data obtained from the numerical simulation, it turned out that this new formula reveals higher accuracy in comparison with residuals.

This monograph is recommended for students and PhD students enrolled in mechanics and materials technology courses, scientists interested in experimental mechanics and engineers dealing with the determination of elasto-plastic material properties from experiments.

Magdalena Gromada
Gennady Mishuris
Andreas Öchsner

Contents

Notation

Variable	Explanation
$D_\varepsilon = \varepsilon - \varepsilon^0$	Deviator of strain tensor
$D_\sigma = \sigma - \sigma^0$	Deviator of stress tensor
F	Tensile force
$I_1(\sigma) = \mathrm{Tr}\sigma = \sigma_r + \sigma_z + \sigma_\theta$	First tensor invariant
$I_2(\sigma) = \frac{1}{2}[\mathrm{Tr}\sigma^2 - \mathrm{Tr}^2\sigma]$	Second tensor invariant
$I_3(\sigma) = \det\sigma$	Third tensor invariant
$J_1(\varepsilon) = 0$	First deviator invariant
$J_2(\varepsilon) = I_2(\varepsilon) + \frac{1}{3}I_1^2(\varepsilon)$	Second deviator invariant
$J_3(\varepsilon) = I_3(\varepsilon) + \frac{1}{3}I_1(\varepsilon)I_2(\varepsilon) + \frac{2}{27}I_1^3(\varepsilon)$	Third deviator invariant
$J_0(\xi)$	Bessel function of zero order
$J_1(\xi)$	Bessel function of first order
K	Elastic bulk modulus
L_0	Initial length of sample
R	Curvature radius of contour of the deformed sample
R_{eH}	Upper yield point
R_{eL}	Lower yield point
R_{m}	Ultimate strength
$R_{\mathrm{p0.2}}$	Conventional yield point
S	Current cross section area of sample
S_0	Initial cross section area of sample
V	Volume of material
a	Current sample radius in the minimum cross section
a_0	Initial radius of sample
k	Yield stress
$\bar{k}$	Average yield stress
$\bar{\mathbf{u}}$	Displacement vector
r	Radial coordinate in current configuration
$r = a_z(z)$	Radius function of the neck cross section at a distance z from the plane of the minimal section
$\bar{\mathbf{v}}$	Velocity vector

(continued)

(continued)

Variable	Explanation
z	Axial coordinate in current configuration
ΔL	Sample elongation
$\Delta \varepsilon_{\text{int}} = (\varepsilon_{\text{int}} - \bar{\varepsilon}_{\text{int}})/\bar{\varepsilon}_{\text{int}}$	Relative increase of the strain intensity
$\Lambda = 1 - a_0/a$	Parameter characterising the stage of plastic strain accumulated in the whole sample
$\Phi(\sigma_{ij})$	Plastic potential
$\Psi = \Psi(J_2(\varepsilon))$	Known function in the deformation theory of plasticity linking the strain and stress deviators
α, β	Parameters describing function of the curvature radius of the longitudinal stress trajectory
$\delta = a/R$	Parameter describing the stage of strain location in the neck surrounding
$\varepsilon = \Delta L/L_0$	Engineering strain (Cauchy's strain)
$\bar{\varepsilon} = \ln \frac{L}{L_0} = \ln(1 + \varepsilon)$	Logarithmic strain (Hencky's strain)
$\boldsymbol{\varepsilon}$	Strain tensor
$\dot{\boldsymbol{\varepsilon}}$	Strain rate tensor
$\bar{\varepsilon}^{el}$	Logarithmic elastic strain
$\varepsilon^0 = \frac{1}{3}I_1(\varepsilon)$	Isotropic component of stain tensor
ε_{int}	Strain intensity (defined by means of the second invariant of the strain deviator)
$\bar{\varepsilon}_{\text{int}}$	Average strain intensity in the minimal section plane
$\bar{\varepsilon}^{p} = \ln(1 + \varepsilon) - (1 - 2\nu)\frac{\sigma_{\text{true}}}{E}$	True (logarithmic) plastic strain up to the moment of neck creation
$\bar{\varepsilon}^{\text{pl}} = 2\ln\left(\frac{a_0}{a}\right)$	True (logarithmic) plastic strain from the moment of neck creation
θ	Circumferential coordinate in current configuration
λ	Factor of proportionality in the constitutive equation of the plastic flow theory determined from the yield condition for each stage of deformation
λ	Lamé's elasticity coefficient
μ	Lamé's elasticity coefficient
ν	Poisson's ratio
ρ	Curvature radius of the principal stress trajectory σ_3
$\boldsymbol{\sigma}$	Stress tensor
$\sigma^0 = \frac{1}{3}I_1(\sigma)$	Isotropic component of stress tensor
$\sigma_0 = F/S_0$	Engineering stress
$\sigma_{\text{true}} = F/S$	True stress (responding the average axial stress in the minimal section plane $\bar{\sigma}_z$); as distinguished from the shearing stress - τ_{rz}
ψ	Slope angle of the tangent of principal stress trajectory σ_3 to the axis z

Chapter 1
Characterisation of the Tensile Test

Abstract In this chapter, the methods utilised to determine the mechanical material properties are presented. In addition, the static tensile test is characterised taking into account the material constants obtained from the engineering stress–strain curve. The plotting of the flow curve of materials is discussed in detail. In addition, the hypotheses mentioned in literature regarding the onset of neck creation and methods of determining the mechanical properties from the tensile test are presented.

Keyword Tensile test · Flow curve · Necking · Material parameters

1.1 Introduction

The dynamic development of materials technology in the sense of elaborating classical and new construction materials requires accurate test execution and evaluation in order to determine their properties [10]. Mechanical tests can be classified according to different criteria dividing them e.g. in static and dynamic tests, carried out at room, higher as well as lower temperature. Static tests comprise tensile, compression, torsion, bending and shearing tests and also complex state of stresses [30]. Dynamic tests contain for example impact and plastometer tests [9, 30]. Huge importance has also the hardness measurement and creep, relaxation, fatigue as well as crack resistance tests [31].

Among all mentioned above methods of determining the mechanical properties of materials, the static tensile test is characterised by many advantages [21]. Some of them can be listed as follows:

- almost homogeneous stress state in the tensile specimen until the moment of the neck appearance,
- possibility of determination of many different mechanical material properties,

M. Gromada et al., *Correction Formulae for the Stress Distribution in Round Tensile Specimens at Neck Presence*, SpringerBriefs in Computational Mechanics, DOI: 10.1007/978-3-642-22134-7_1, © Magdalena Gromada 2011

- possibility of observation of the elongation process from the beginning of the sample loading until its destruction and the qualitative and quantitative estimation of the process,
- easiness of test realisation.

The tensile test, thanks to its advantages, is commonly known as the basic static resistance test, with the widest application in all fields of material testing [21]. Presently, in time of rapid development of modern materials and new calculation method utilised in the modelling process, the weakest chain is the precise evaluation of material properties. Till now, there are classical formulae used for the yield stress determination after neck creation in the sample which were derived in the 1940 and 1950s. However, these formulae are applied without any estimation of their accuracy in the material parameters determination. The progressive development of measurement methods allows nowadays the exact determination of measurable and additional properties, which was earlier impossible. Therefore, the aim of this monograph is the analysis of the current knowledge state, estimation of the accuracy of all available classical solutions and the presentation of new formulae which allow to accurately reconstruct the plastic properties of materials.

1.2 Determination of the Flow Curve

The procedure to determine basic mechanical materials properties from the tensile test is specified for example in standards ASTM E8/E8M-09 [36], EN 10002-1 [37], ISO 6892-1 [38]. These standards contain appropriate definitions, give sample dimensions and conditions for the test execution.

The cross section of samples for testing can be round, square or ring-shaped and in particular cases specimens can have also another shape [37]. However, in this monograph, the consideration is restricted to the tensile test of axisymmetrical samples and the obtained results do not take into account the influence of the sample shape. Figure 1.1 presents an initial and uniform deformed axisymmetrical sample with the used cylindrical coordinate system.

From the tensile test, the so called tension curve, which represents the relation of the sample elongation ΔL to the value of the tensile force F, can be obtained. However, this diagram does not have a wide application in practice because it makes impossible to compare the obtained results for samples of different cross sections. In order to eliminate the influence of the sample dimensions on the graph shape, the values of engineering strain and stress are calculated and graphically represented. Two different measures of strain are commonly used. When the sample elongation ΔL is referred to the initial measuring length L_0, the engineering strain (Cauchy's) ε [24] is obtained:

Fig. 1.1 Initial and uniform deformed tensile sample with cylindrical coordinate system

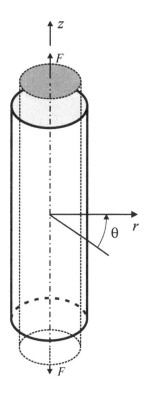

$$\varepsilon = \frac{\Delta L}{L_0}. \tag{1.1}$$

If we sum up the strain increases and refer it to the temporary sample length, the logarithmic strain (Hencky's) $\bar{\varepsilon}$ [24] is obtained:

$$\bar{\varepsilon} = \ln \frac{L}{L_0} = \ln(1 + \varepsilon). \tag{1.2}$$

It should be noticed here that determined in this way strains ε_z refer to the axial or loading direction. Whereas, the measure of the engineering stress σ_0 is the ratio of the tensile force F and the initial cross section area of the sample S_0:

$$\sigma_0 = \frac{F}{S_0}. \tag{1.3}$$

Such a procedure assumes uniform sample deformation on the whole of its measuring length and does not take into consideration the local change of cross section area of the specimen (cf. Fig. 1.1).

For this reason the diagram of stress as a function of strain (Fig. 1.2) plotted on the base of formulae (1.2) and (1.3) is called the *engineering* stress–strain curve of the material.

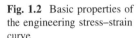

Fig. 1.2 Basic properties of
the engineering stress–strain
curve

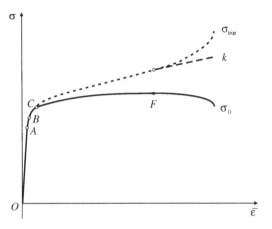

Among two measures of strain according to Eqs. 1.1 and 1.2 for plotting the stress–strain curve, the logarithmic strain $\bar{\varepsilon}$ was chosen for the sake that it is linked with the known formula for the true (logarithmic) plastic strain. Moreover, its application is recommended in many publications [18, 24].

Let us discuss the basic features of the engineering stress–strain curve in coordinates σ_0 and $\bar{\varepsilon}$ as shown in Fig. 1.2. At the beginning, the segment OA of this diagram is a straight-line. The end point A of this segment determines a stress called the limit of proportionality. Strains in the range OA are pure elastic and they disappear after the loading is removed. In a small range behind the OA segment, the strain reveals an elastic character. However, the stress from which the first plastic strain performs is greater than the limit of proportionality.

This value is determined by the point B in Fig. 1.2, whose ordinate refers to the value of stress called the elastic limit. Taking into account that location of this point is very difficult to be found, the notion of the conventional elastic limit interpreted by such a value of stress causing a plastic strain of 0.01 or 0.02% is introduced. In the case of materials which do not exhibit a clearly yield point (Fig. 1.2) in engineering practice the notation of a conventional yield limit $R_{p0.2}$ is applied which equals to a stress producing a plastic strain of 0.2% (point C in Fig. 1.2).

If the engineering stress–strain curve has a character as presented in Fig. 1.3, the determination of the yield limit does not produce any problem.

Subsequently, we distinguish the so-called upper yield point R_{eH}, which is defined as the value of stress at the moment when the first decrease of force appears and the lower yield point R_{eL}, namely the lowest stress during flow with the omission of possible initial temporary effect [37]. In order to generate further strains, the load must be increased. This phenomenon is called hardening of the material. The slope coefficient of this part of the diagram denoted by points CD is called the hardening coefficient. To better explain the hardening effect, the deformed sample can be unloaded and again loaded to the point D (Fig. 1.4).

Fig. 1.3 Fragment of the
engineering stress–strain
curve with distinction of yield
limits

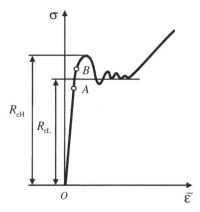

Fig. 1.4 Engineering
stress–strain curve with
unloading

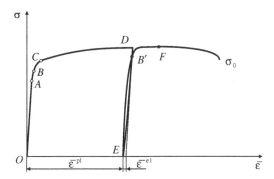

During removal of the loading force from point D to E, only the elastic part of the strain $\bar{\varepsilon}^{\text{el}}$ disappears. However, part of the plastic strain $\bar{\varepsilon}^{\text{pl}}$ remains. At redirect loading the strains are elastic until point B' is reached. This point is placed above the point B. It follows from this that the material hardened as a result of plastic deformation. At further loading, the plastic strains will again increase. It should be noticed that after a short segment of great curvature, the diagram returns to the initial part before unloading. If the tension load after point D continues further without unloading, subsequently both diagrams practically overlap. An important feature of the material from the engineering point of view is the ultimate strength R_{m} determined as the stress referring to the maximal tensile force (point F in the Fig. 1.2 and 1.4).

Obviously, the value of the engineering stress in the tensile test does not refer to the true stress in the sample because the cross section area decreases from the moment of plastic strain formation in the sample. This process is especially visible in the final stage of tension when the sample is subject to the highly local contraction called necking. This phenomenon creates the characteristic decrease of the tensile force, namely the decrease in engineering stress. If we take as a measure of stress the ratio of the current value of tensile force and the current minimal cross section area of the sample, the true stress is obtained as:

$$\sigma_{\text{true}} = \frac{F}{S}, \tag{1.4}$$

which increases up to the moment of fracture (Fig. 1.2). Up to the neck appearance, the true stress is practically equal to the engineering stress, e.g. [21, 24]. Let us link in the following the true and engineering stress with the initial and current cross section areas as:

$$\sigma_{\text{true}} = \sigma_0 \frac{S_0}{S}. \tag{1.5}$$

During the derivation of the formula for the true stress it is assumed in all engineering textbooks that the sample volume does not change during the tensile test. As a result of this assumption, one gets:

$$\sigma_{\text{true}} = \sigma_0 \frac{L}{L_0}. \tag{1.6}$$

Taking advantage of formula (1.1), the final relation (see also [23, 35]) is obtained as:

$$\sigma_{\text{true}} = \sigma_0(1 + \varepsilon). \tag{1.7}$$

Following an analogous procedure, but taking the change of the sample volume into account by consideration of the elastic strain component, it can be shown that

$$\sigma_{\text{true}} = \frac{\sigma_0(1 + \varepsilon)}{1 + \frac{\sigma_0}{K}}, \tag{1.8}$$

where K is the elastic bulk modulus satisfying the condition $E \leq K < \infty$ while E is Young's modulus.

Let us notice that, for example, for steel 30 HGS [24] at the moment of the plastic strain appearance, the ratio σ_0/K is of the order of 0.6%, while for the ultimate strength it reaches 0.9%. Therefore, this value cannot be omitted in the further considerations if a higher accuracy should be obtained.

The true stress–strain curve up to the moment of neck creation is estimated by formula (1.7) or (1.8) and from the moment of localised contraction according to formula (1.4). This curve reveals at the end a strong increase, which is the result of the neck appearance and a triaxial state of stress. The neck creation disturbs the uniformity of the uniaxial state of stress. The stress trajectories near the neck contour cannot proceed parallel to the sample axis as well as radial and circumferential tension stresses will appear in addition in the minimal cross section.

Therefore, the determination of the true stress–strain curve must consider this triaxial stress state from the moment of neck creation. As a result of a such correction, the curve $k = k(\varepsilon_{\text{int}})$ is obtained which characterises the material properties during deformation, cf. plot on graph 1.2. However, instead of the strain intensity value ε_{int} (see Eq. 2.18), $\bar{\varepsilon}$ was consistently assumed from the moment of neck creation which enables to compare the engineering and the true curve with the corrected k.

In engineering practice, the flow curves are determined from the moment of plastic strain appearance in the sample (see [24, 27]). Subsequently, the independent variable is the plastic true strain but this value can be obtained in two ways. Up to the moment of neck creation one can subtract from the logarithmic strain (1.2) the elastic strain to obtain:

$$\bar{\varepsilon}^{p} = \ln(1 + \varepsilon) - (1 - 2\nu)\frac{\sigma_{\text{true}}}{E}, \tag{1.9}$$

where ν is Poisson's ratio and E is Young's modulus.

If the value of engineering strain at which the neck occurs does not exceed 10%, the engineering strain can be approximately considered for the logarithmic strain and this simplification does not change the graph in direction of the ordinate. This remark also holds in regards to formula (1.9). In paper [32] there is an incorrect form of relation (1.9), where the effect of Poisson's ratio ν is not taken into consideration.

From the moment of neck creation, the formula for the true plastic strain also utilised by Bridgman can be applied. It was derived under omission of the elastic strain and under the assumption of material incompressibility. Determining the length ratio L/L_0 from the condition of constant volume, i.e. $V = S_0 L_0 = SL$, and inserting it into Eq. 1.2, one obtains after simplification:

$$\bar{\varepsilon}^{\text{pl}} = \ln\left(\frac{S_0}{S}\right). \tag{1.10}$$

The formula for the initial cross section area using the initial radius a_0 can be written as:

$$S_0 = \pi a_0^2, \tag{1.11}$$

while for the present cross section area with the radius of the sample after deformation a, has the form:

$$S = \pi a^2. \tag{1.12}$$

Inserting relations (1.11) and (1.12) into (1.10), one finally gets:

$$\bar{\varepsilon}^{\text{pl}} = 2\ln\left(\frac{a_0}{a}\right). \tag{1.13}$$

This formula is known in the literature (see for instance [1, 4, 18, 24, 27]) as a relation for the true (logarithmic) plastic strain.

In deformation theory of plasticity, the true stress–strain curve is the diagram of the stress intensity k as function of the strain intensity ε_{int}. Determining the strain intensity from formula (2.18) in Sect. 2.1.3 under the assumption of material incompressibility (it can be also assumed at small plastic strains, as it follows from discussion after formula (1.8), that the consideration of the material volume change leads to an error of less than 1%), one obtains that the strain intensity is equal to the axial strain $\varepsilon_{int} = \varepsilon_z$ (see also [11, 24]). However to prove such a property from the moment of neck appearance, it is additionally assumed that the radial and the circumferential strains are equal [24]. Subsequently, the strain intensity is also equal to the axial strain ($\varepsilon_{int} = \varepsilon_z$). Let us notice that in the context of plastic flow theory it is assumed that $k = k(\bar{\varepsilon}^{pl})$ and thus, the diagrams $\sigma_{true} - \varepsilon_{int}$ must be further corrected. This is possible because the elastic and plastic strains can be separated in the theory of small deformations. The first one satisfies Hooke's law while the second one, i.e. the plasticity law, must be determined in other ways in deformation as well as plastic flow theories of plasticity. Taking this into consideration, the curves can be corrected, up to the moment of neck creation by formula (1.9), and from the moment of localised contraction by relation (1.13). Based on the fact that the simplification of the radial and circumferential strains equality can be considered as questionable, the verification of the formula (1.13) will be performed in Sect. 3.3 utilising outcomes obtained from numerical simulation. However, as it was underlined above, the true stress is determined up to the moment of neck creation from formula (1.7) and subsequently the correction is made in order to take into account the triaxial stress state.

In literature, one can very often find a correction coefficient [3, 8] to determine the value of which the true stress must be multiplied in order to obtain the yield stress. The correction coefficient is the inverse value of the average normalised axial stress.

The issue to estimate a complex stress state under the neck presence and to obtain the relation $k = k(\varepsilon_{int})$ from the measurements was considered by many scientists (see Sect. 1.4). However, up to now it is common to use the formulae obtained in the 1940 and 1950s. In Sect. 3.4 it will turn out that their application for the yield stress estimation for ideal plastic materials can generate even ten percent of error in comparison with data assumed in the numerical simulation. Therefore, new formulae for the yield stress will be presented in this monograph, which can better approximate the flow curve of a material.

1.3 Phenomenon of the Neck Creation

The review of the tensile test requires to be also mentioned that it is not easy to determine the onset of neck creation. According to the engineering criterion, the onset of neck creation for the majority of materials occurs when the tensile force reaches the maximum value [13, 16, 24, 40]. This hypothesis, known as the Considère criterion, is a result of experimental tests conducted on samples made from iron and steel. It was observed that at the moment of neck initiation, the sample is in an unstable equilibrium, i.e. the force reaches a maximum value and subsequently decreases. It was finally concluded from analytical together with experimental results that the point at maximum load corresponds to the onset of neck creation. This hypothesis was also confirmed recently by Havner in paper [14], where the author carried out an analytical analysis in order to examine the rightness of the Considère criterion for the onset of neck creation in the case of material inhomogeneity. The obtained results form the theoretical basis for the Considère observation that the maximum total load corresponds to the onset of neck creation. Paper [14] concludes with consideration regarding of geometrical irregularities or imperfections in ideal homogeneous materials which is more often examined in literature. As a result of the performed analysis, the author admits that the force stays constant during a short time when the neck appears.

Many scientists applied the finite element method to determine the onset of neck creation. In paper [19], a solution of the numerical simulation satisfying accurately in engineering sense the Considère criterion was obtained for the first time. Within the scope of these simulations, round samples were utilised without consideration of any imperfection.

In the following, we will present the classical solution bearing in mind that up to the moment of neck creation a homogeneous state exists, where the current cross section area of the sample S and the true stress σ_{true} undergo a change. At the point where the tensile force takes its maximal value, one gets:

$$dF = \sigma_{\text{true}}\, dS + S\, d\sigma_{\text{true}} = 0. \tag{1.14}$$

Assuming that the plastic strain does not cause a change of the material volume V, one obtains:

$$dV = S dL + L\, dS = 0. \tag{1.15}$$

Comparing relations (1.14) and (1.15), one gets:

$$\frac{d\sigma_{rz}}{\sigma_{rz}} = \frac{dL}{L} = d\bar{\varepsilon}, \tag{1.16}$$

which leads to the known relation given by A. Considère:

$$\frac{d\sigma_{\text{true}}}{d\bar{\varepsilon}} = \sigma_{\text{true}}. \tag{1.17}$$

Last equation can be interpreted as the condition of stability loss under tensile load which occurs when the hardening coefficient $d\sigma_{\text{true}}/d\bar{\varepsilon}$ reaches the value of the current true stress. In textbook [5], as a result of more general derivations which took into account the sample dimensions, the following condition for the onset of neck creation was obtained in the form:

$$\frac{1}{\sigma_{\text{true}}} \frac{d\sigma_{\text{true}}}{d\bar{\varepsilon}} \approx 1 - \frac{1}{2}\left(\frac{\pi a}{L}\right)^2. \tag{1.18}$$

Fig. 1.5 Graphical
determination of the point of
neck creation

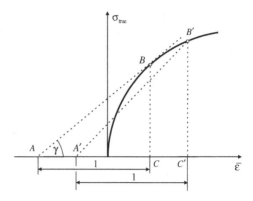

Obviously, in the case of appropriate long specimens, this condition reduces to the relationship of Eq. 1.17.

In addition, there is a graphical method for the determination of the onset of neck creation [16] based on formula (1.17). On the empirically estimated general diagram of the relation between stress and strain given in coordinates σ_{true} and $\bar{\varepsilon}$, a segment $A'C'$ of the length of 1, which crosses the coordinate origin can be drawn. At point C' a vertical line to intersect the graph to obtain point B' can be drawn. Shifting segment $A'C'$ to the right or left of the graph, a triangle with corner points $A'B'C'$ can be determined, so that the line $A'B'$ is tangential at point B' to the graph. In such a manner, the only solution in the form of the triangle ABC can be got. This is presented in the Fig. 1.5 by point B which indicates the onset of neck creation.

Determining $\tan\gamma$ from triangle ABC, one gets $\tan\gamma = BC/AC = \sigma_{\text{true}}/1$, which coincides with condition (1.17).

However the author of textbook [40] quotes empirical examples in which it was stated that the onset of neck creation occurs considerably beyond the maximal force which was treated as exceptional situations. Furthermore, analytical analysis and the numerical simulation of the process of neck creation were carried out and they exhibited that the length to diameter ratio of the sample has influence on the beginning of the neck formation [40, 46]. This thesis was also confirmed in publication [17], were the authors theoretically treated the bifurcation problem manifesting the onset of neck creation. The obtained results testify that the true stress at bifurcation is greater than the stress at which the maximal load is achieved by a value which depends on the radius to length ratio of the sample, the ratio of the elastic shear modulus to the tangent modulus and the derivative of the tangent modulus with respect to the stress. Needleman [26] performed numerical simulations of samples, whose ends were placed in stiff grips. The results confirm that in the case of a specimen of small length and large diameter, the beginning of the neck creation is shifted to a greater value of obtained strain. Opposed to the previous work, the author of paper [46] tries to determine the onset of neck creation without the necessity of experiments for the known material properties. This work leads to an analogous relation as obtained in Eq. 1.17 but for $k = k(\varepsilon_{\text{int}})$.

However, this method is very complicated and it assumes that the shape of the appearing neck is known.

In the process of the material flow curve determination, it is not necessary to accurately estimate the exact onset of neck creation. At the initial stage of the contraction, both the true strain and the ratio of the current sample radius to the current curvature radius of the neck contour—a/R, being the argument of all classical formulae, take values only slightly greater than zero. Thus, it is enough to estimate the moment of neck formation based on experimental observations.

1.4 Analysis of the State of Knowledge Regarding the Mechanical Properties Determination from the Tensile Test at the Neck Presence

The great significance of the analysis of the stress distribution in the neck of axisymmetrical tensile specimens stimulated papers in which the authors presented detailed formulae derivations for the yield stress from the moment of neck formation. The first solutions of this problem were obtained by Bridgman [3], Siebel [34], Davidenkov-Spiridonova [7] and subsequently also Szczepiński [8] as well as Malinin and Petrosjan [24]. These formulae are content of many textbooks in the field of material properties determination [21, 24, 39, 40] and the theory of plasticity [15, 20]. Detailed derivations of the classical formulae by Bridgman, Siebel, Davidenkov-Spiridonova, Szczepiński as well as Malinin and Petrosjan will be presented in Chap. 2. However, in this review, a few fundamental matters will be discussed from Bridgman book [3], where he introduced the formula for the correction coefficient of the flow curve from the moment of neck appearance both for axisymmetric samples and rectangle cross section specimens. Obviously, these formulae for the yield stress in both cases have the same form but it should be noticed that the relation for the average axial stress differs from each other for different cross sections.

The independent variable determined by Bridgman for the flow curve from the moment of neck creation is the true strain (see Sect. 1.2). The formula derived by Bridgman for the yield stress is a function of the ratio of the current sample radius a and the current curvature radius of the neck contour R. Taking into account that the radius R is the value which is very difficult to determine in experimental practice, he linked, on the basis of experimental results, the ratio a/R with the true plastic strain. This relationship was verified in other papers [4, 25], where it was unanimously stated that this formula does not involve enough accuracy to approximate the ratio a/R and for this reason it is not recommended for application. Therefore, the Bridgman endeavours to derive a formula of only one variable, which is very easy for determination, did not succeed.

In a short time after the publication of the classical formulae for the determination of the average normalised axial stress in the minimal section plane, attempts were made to verify the application based on data obtained from experiments [25]. Since the yield stress cannot be obtained from experiments, the authors compared the value of the true stress with the outcomes of the classical formulae. It turned out that the Bridgman formula exhibits constant accuracy during the process of the whole tensile test. However, results obtained from the Daviedenkov-Spiridonova formula for a value of a/R greater than 0.6 were not so accurate. On this basis the conclusion was drawn that the application of the Bridgman formula causes outcomes closer to the true solution. Obviously, this does not have to be true and requires further research in each separate case.

During the derivation of the formulae for the yield stress, Bridgman, Siebel, Davidenkov-Spiridonova, Szczepiński as well as Malinin and Petrosjan utilised certain simplifying assumptions. From the moment of computer program dissemination for the numerical simulation based on the finite element method, many papers were written [6, 26, 27, 33, 44] with the attempt to verify the classical formulae and their assumptions: herein on the equality of the radial and circumferential stresses as well as the constancy of the yield stress in the plane of the minimal section. These two assumptions as a result of theoretical analysis were recognised as questionable in many textbooks e.g. [8, 15, 20]. The statement that the simplifications have a poor foundation gave rise for new research to eliminate them in order to improve the existing formulae.

A few authors [6, 22] utilised the numerical simulation of the tensile test to estimate the error involved in the application of the classical formulae. They admit the considerable discrepancy between results obtained from numerical simulation and the ones obtained from the classical formulae. Such analysis is also presented in Sect. 3.4 of this monograph.

It must be noted that the first information on a better approximation of the true solution by the Davidenkov-Spiridonova formula compared to the Bridgman approach was given by Jasieński in paper [18], in which the author quoted research results by C. Rossard, P. Blam and F. A. Hodierne. Despite this information, the Bridgman formula is up to now the most often applied approximation in engineering practice for the flow curve determination from the beginning of the neck creation [4, 22, 23, 28, 41, 44, 45]. Additionally, authors of some papers utilise this flow curve e.g. for the creation of fracture models for plastic materials [29, 32, 45]. Obviously, such solution generates errors at the stage of input data generation.

In many publications [1, 32, 42, 43, 45] the Bridgman formula, which is derived for smooth axisymmetric samples, is applied for the determination of the flow curve of notched axisymmetric tensile specimens. However, it is impossible to find in the literature enough confirmation for the correctness of such a procedure. Only in paper [42], the author specifies the assumptions made by Bridgman during the formula derivation for smooth axisymmetric samples and claims that to a first approximation this formula can be applied also for notched axisymmetric specimens.

The measurement difficulties of the profile of deformed samples and the problems in the determination of the curvature radius of the neck contour R were noticed by many researchers [4, 12, 18, 23, 40, 41, 44] and it was even mentioned by Bridgman. Simultaneously, it is known that the error involved in the

R determination additionally influences the value of the yield stress determined from the optionally chosen formula. Therefore, a few publications were written in which the neck profile was linked with different values which are easy to measure during the experiment. Paper [18] presents a relation derived by Krupkowski and Wantuchowski for the neck shape as a function of the stage of contraction. However in paper [12], a formula was derived which approximates the neck profile, based on two experimental parameters describing the shape of the specimen. Paper [41] presents the estimation of the theoretical profile of deformed specimens, which indeed coincides with the shape obtained from experiments. However, the modelling was made only for ideal plastic materials. Szczepiński proposed in a book [8], in addition to the formula derived in accordance with the classical assumptions for the determination of the average normalised axial stress distribution in the minimal section plane, to apply the slip line method which will eliminate the necessity of the radius R calculation. However, La Rosa [22] obtained a law which does not require the radius R knowledge while it depends only on the deformation of the neck according to which the true stress should be corrected. However, the flow curve obtained as a result of such a law possesses the same accuracy as the values determined by Bridgman's formula. The reason is that this relation is derived on the basis of data obtained from the numerical simulation of the flow curve determined from the Bridgman formula.

The first papers regarding the improvement of the classical formulae appeared immediately after the publication of a classical approximation. Among others, Truszkowski [18] dealt with the derivation of the new formula for the average normalised axial stress. In his consideration, he based the strain analysis on the principle of the minimal work of deformation and he utilised the above mentioned formula by Krupkowski and Wantuchowski which enables the determination of the neck profile of the axisymmetric sample. On the basis of experimental tests, he derived a relation between the average hardness according to Meyer and the average yield stress. However, the derived formula requires additionally a hardness measurement. Therefore, the verification of the formula derived by Truszkowski is not so easy as in the case of the classical approximations.

Paper [24] quotes a solution proposed by Malinin and Petrosjan. In their consideration, the authors utilised two equilibrium equations in the surrounding of the minimal section and the simplifying assumptions on the equality of the radial and circumferential stresses as well as the constancy of the yield stress in the minimal section. As a result of the executed analytical calculations, a formula for the stress intensity was obtained, which application requires the knowledge of Bessel function. However, as it will be presented in Chap. 2, the got solution can be written in a simple form for application by engineers.

In a separate group of publications [2, 6, 26, 27, 33, 44], the authors created analytical models on the basis of the Prandtl-Reuss plastic flow theory with the Huber-Mises yield condition in Lagrange coordinates with nonlinear equilibrium equations. The solution of this approach was determined by application of the finite difference method. As a result, they obtained the flow curve but they did not analyse the stress and strain states in the neck region. However, these calculations were quite complicated and did not deliver the solution for a greater stage of material deformation.

During the last years, the yield stress determination for axisymmetrical and plane tensile samples got again a great interest. Among other issues, Ling applied in paper [23] the method of weighed average for the estimation of the flow curve from flat samples. However, this method requires the knowledge of numerical simulation results of the upper and lower limit of the flow curve and is not useful in practice. In a same manner, the method presented in publication [19] requires except the experimental tests of axisymmetrical samples a numerical simulation of the tensile test. The authors claim that the application of a very complicated iterative algorithm allows to obtain high accuracy of the solution but the manner of calculation execution is very laborious.

The above collected information and concluded points justify the necessity of further analyses and presumable improvement of the classical formulae for the yield stress. It is also justified by the fact that in these days of modern computers where it is aimed to achieve calculation accuracies in the range of percent fractions and measurements which are conducted by means of very sensitive extensometers, e.g. based on lasers, errors in the determination of materials properties following from imperfections of applied formulae are no longer acceptable. Simultaneously, as it was presented in the literature review, the Bridgman formula is up to now commonly used not only for the flow curves estimation but also e.g. for the generation of input data for creation of fracture models of plastic materials [29, 32, 45]. In some of the mentioned publications, the simplifying assumptions applied in the classical approaches are verified by means of numerical simulations. It was questioned among other the equality of the radial and circumferential stresses and the constancy of the yield stress in the minimal section plane. However, the assumption of the form of the formula for the curvature radius of the longitudinal stress trajectories, which is essential for the Bridgman and Siebel-Davidenkov-Spiridonova formulae, is until now not verified even using data from numerical simulations. In addition, the order of error connected with the application of these assumptions was not estimated. For this sake it is purposeful to verify all the three above mentioned questionable assumptions: namely the equality of the radial and circumferential stresses, the constancy of the yield stress in the minimum section plane and the form of formula for the curvature radius of the longitudinal stress trajectories. However, it is not enough to state the untruth of assumptions because almost every assumption cannot be true at the end. The error following from the application of particular simplifications should be estimated. This will allow to eliminate during a new formula derivation all questionable assumptions or, at least, the assumptions which generate the greatest errors.

Despite the fact that many attempts were made in order to determine the approximating function of the neck shape and linking the radius R with values easily measurable during experiments, none of them brought the expected results. Therefore, every effort should be done so that the determination accuracy of the curvature radius of the neck contour R is the highest possible.

One more conclusion follows from the literature analysis. In many publications, the independent variable of the flow curve from the moment of the neck appearance is the true plastic strain instead of the strain intensity. However, it is difficult to find the verification of this formula and to estimate its accuracy. Therefore, the correctness of the application of the formula for the true plastic strain instead of the strain intensity is checked in Sect. 3.3 which is based on numerical results.

References

1. M. Alves, N. Jones, Influence of hydrostatic stress on failure of axisymmetric notched specimens. J. Mech. Phys. Solids **47**, 643–667 (1999)
2. V.G Bazhenov, A.I Kibec, P.V Laptev et al. *Experimental and Theoretical Investigation of the Limiting State of Elasto-Plastic Specimen with Different Cross Section Under Tensile Test (in Russian). Problems of Mechanics* (Fizmatlit, Moscow, 2003), pp. 115–122
3. P.W. Bridgman, *Studies in Large Plastic Flow and Fracture with Special Emphasis on the Effects of Hydrostatic Pressure* (Harvard University press, Cambridge, 1964)
4. E.E. Cabezas, D.J. Celentano, Experimental and numerical analysis of the tensile test using sheet specimens. Finite Elem. Anal. Des. **40**, 555–575 (2004)
5. J. Chakrabarty, *Applied Plasticity* (Springer, New York, 2000)
6. W.H. Chen, Necking of a bar. Int. J. Solids Struct. **7**, 685–717 (1971)
7. N.N. Davidenkov, N.I. Spiridonova, Mechanical methods of testing. Analysis of the state of stress in the neck of a tension test specimen. Proc. Am. Soc. Test Mater. **46**, 1147–1158 (1947)
8. L. Dietrich, J. Miastkowski, W. Szczepiński, *Limiting Capacity of the Construction Elements (in Polish)* (PWN, Warsaw, 1970)
9. L.A. Dobrzański, R. Nowosielski, *Methods of Metals and Alloys Testing. Mechanical and Physical Properties Testing (in Polish)* (Publishers of Silesian University of Technology, Gliwice, 1986)
10. Z. Dylg, A. Jakubowicz, Z. Orłoś, *Strength of materials (in Polish)* (PWN, Warsaw, 1996)
11. Z. Gabryszewski, J. Gronostajski, *Mechanics of Process of Plastic Forming (in Polish)* (PWN, Warsaw, 1991)
12. W. Gaudig, K. Bothe, A.K. Bhaduri et al., Determination of the geometric profile and stress/strain state in the necked region during inelastic deformation at elevated temperatures using a non-contact measurement technique. Test Evaluat. **24**, 161–167 (1996)
13. M.W. Grabski, *Structural Superplasticity of Metals (in Polish)* (Silesian Publishers, Katowice, 1973)
14. K.S. Havner, On the onset of necking in the tensile test. Int. J. Plast. **20**, 965–978 (2004)
15. R. Hill, *The Mathematical Theory of Plasticity* (Clarendon Press, Oxford, 1950)
16. O. Hoffman, G. Sachs, *Introduction to the Plasticity Theory (in Polish)* (PWN, Warsaw, 1959)
17. J.W. Hutchinson, J.P. Miles, Bifurcation analysis of the onset of necking in an elastic/plastic cylinder under uniaxial tension. J. Mech. Phys. Solids **22**, 61–71 (1974)
18. Z. Jasieński, Influence of the deformation irregularity on the relation of the specific stress as stage of deformation in neck of metal tensile specimen (in Polish). Arch. Metall. Mater. **10**, 189–239 (1965)
19. M. Joun, J.G. Eom, M.C. Lee, A new method for acquiring true stress-strain curves over a large range of strains using a tensile test and finite element method. Mech. Mater. **40**, 586–593 (2008)
20. L.M. Kachanov, *Foundations of the Theory of Plasticity* (Mir Publishers, Moscow, 1974)

21. S. Katarzyński, S. Kocańda, M. Zakrzewski, *Investigation of Mechanical Properties of Metals (in Polish)* (WNT, Warsaw, 1967)
22. G. La Rosa, G. Mirone, A. Risitano, Postnecking elastoplastic characterisation: degree of approximation in the Bridgman method and properties of the flow-stress/true-stress ratio. Metall. Mater. Trans. **34**, 615–624 (2003)
23. Y. Ling, Uniaxial true stress-strain after necking. AMP J. Tech. **5**, 37–48 (1996)
24. N.N. Malinin, J. Rżysko, *Mechanics of Materials (in Polish)* (PWN, Warsaw, 1981)
25. E.R. Marshall, M.C. Shaw, The determination of flow stress from a tensile specimen. Trans. Am. Soc. Metal **44**, 705–725 (1952)
26. A. Needleman, A numerical study of necking in circular cylindrical bars. J. Mech. Phys. Solids **20**, 111–127 (1972)
27. D.M. Norris, B. Moran Jr, J.K. Scudder et al., A computer simulation of the tension test. J. Mech. Phys. Solids **26**, 1–19 (1978)
28. A. Öchsner, *Experimental and Numerical Investigation of the Elasto-Plastic Behaviour of Cellular Model Materials (in German)* (VDI Verlag GmbH, Düsseldorf, 2003)
29. A. Öchsner, J. Gegner, W. Winter et al., Experimental and numerical investigations of ductile damage in aluminium alloys. Mater. Sci. Eng. A-Struct **318**, 328–333 (2001)
30. K. Przybyłowicz, *Metal Science (in Polish)* (WNT, Warsaw, 2003)
31. K. Przybyłowicz, J. Przybyłowicz, *Review of Materials Science. Methods of Testing The Metal Materials (in Polish)* (Publishers of Świętokrzyska University of Technology, Kielce, 2002)
32. L. Ramser, W. Winter, G. Kuhn, in *Ductile Damage in Aluminium Alloys*. Workshop on Advanced Computational Engineering Mechanics, Maribor, 2003
33. M. Saje, Necking of a cylindrical bar in tension. Int. J. Solids Struct. **15**, 731–742 (1979)
34. E. Siebel, S. Schwaigerer, On the mechanics of the tensile test (in German). Arch. Eisenhuttenwes **19**, 145–152 (1948)
35. M. Skorupa, Engineering and True Flow Curve (in Polish). http://kwitm.imir.agh.edu.pl/mskorupa/pdf/02.pdf
36. Standard ASTM E8/E8 M-09 Standard Test Methods for Tension Testing of Metallic Materials
37. Standard EN 10002 –1 Metallic materials – Tensile testing – Part 1: Method of test at ambient temperature
38. Standard ISO 6892-1Metallic materials-Tensile testing-Part 1: Method of test at room temperature
39. N.A. Shapashnikov, *Mechanical Tests of Metals (in Russian)* (Mashgiz, Moscow, 1954)
40. W. Szczepiński, *Experimental Methods of Solid Mechanics (in Polish)* (PWN, Warsaw, 1984)
41. P.F. Thomason, An analysis of necking in axi-symmetric tension specimens. Int. J. Mech. Sci. **11**, 481–490 (1969)
42. A. Valiente, On Bridgman's stress solution for a tensile neck applied to axisymmetrical blunt notched tension bars. J. Appl. Mech. **68**, 412–419 (2001)
43. A. Valiente, J. Lapena, Measurement of the yield and tensile strengths of neutron-irradiated and post-irradiation recovered vessel steels with notched specimens. Nucl. Eng. Des. **167**, 11–22 (1996)
44. K.S. Zhang, Z.H. Li, Numerical analysis of the stress-strain curve and fracture initiation for ductile material. Eng. Fract. Mech. **49**, 235–241 (1994)
45. K.S. Zhang, C.Q. Zheng, Analysis of large deformation and fracture of axisymmetric tensile specimens. Eng. Fract. Mech. **39**, 851–857 (1991)
46. A.M. Zhukov, On the problem of the neck appearance in the tensile test (in Russian). Collect Engineers **2**, 4–51 (1949)

Chapter 2
Stress Distribution in the Sample Neck during the Tensile Testing

Abstract In this chapter, the considered issue is presented on the base of two different models of plasticity theory and the detailed derivation of formulae for the average normalised axial stress in the minimal section plane is shown based on different classical approaches. For the sake of fact that all classical derivations are presented in so-called Euler coordinates, the independent variables are coordinates in the current configuration of the point in the space, in which a material particle is located at the considered moment of time. Let us notice that in this coordinate system for the given deformation, an additional problem is the determination of the boundary positions along which the boundary conditions on the neck contour are estimated since they are continuous changing for each stage of deformation. However, this inconvenience can be recompensed by linear equilibrium equations [8].

Keywords Stress distribution in the sample with neck · Classical and new approaches · Deformation theory · Flow theory

2.1 Problem Description based on Deformation Theory of Plasticity

2.1.1 Consideration of the Axial Symmetry

The considered problem is axisymmetric, which follows from the neck shape and the manner of its loading. Let us notice that during the experiment at certain end stages of deformation, the deviation from perfect axial symmetry can take place. This is connected with imperfections during the sample manufacturing as well as with the fact that the state of loading is not till the end perfectly axisymmetric. However, it is always assumed in the analytical modelling that the sample is

M. Gromada et al., *Correction Formulae for the Stress Distribution in Round Tensile Specimens at Neck Presence*, SpringerBriefs in Computational Mechanics, DOI: 10.1007/978-3-642-22134-7_2, © Magdalena Gromada 2011

perfect both in regard to material and shape. These assumptions hold through this monograph.

To sum up, the assumption that the issue is axisymmetric originates from the fact that all values (displacements, strains, stresses) depend solely on two cylindrical coordinates r as well as z and do not depend on the third coordinate θ (Fig. 1.1):

$$\bar{\mathbf{u}} = \bar{\mathbf{u}}(r, z), \tag{2.1}$$

$$\underline{\varepsilon} = \underline{\varepsilon}(r, z), \tag{2.2}$$

$$\underline{\sigma} = \underline{\sigma}(r, z), \tag{2.3}$$

and the vector and tensor components in direction of the variable θ are equal to zero:

$$u_\theta = 0, \tag{2.4}$$

$$\varepsilon_{r\theta} = 0, \quad \varepsilon_{z\theta} = 0, \tag{2.5}$$

$$\tau_{r\theta} \equiv 0, \quad \tau_{z\theta} \equiv 0. \tag{2.6}$$

Due to the sample symmetry as well as symmetry of the load and strain states with respect to the plane $z = 0$, one can state that all functions must be even or odd in regard to this variable. Namely, u_z has to be odd in respect of the z variable while u_r must be even. From Eq. 2.24 it follows that ε_{rz} is an odd function with respect to the z variable and thus, $\varepsilon_{rz}(r, 0) = 0$. As it follows from the equation of deformation theory of plasticity (2.27) and in addition from the physical sense that the stresses behave very similar, therefore τ_{rz} is an odd function with respect to the z variable. Thus:

$$\tau_{rz}(r, z) = 0 \quad \text{when } z = 0. \tag{2.7}$$

In a similar manner, the analysis of the stress state can be carried out for the component τ_{rz} and the axis $r = 0$ to finally obtain:

$$\tau_{rz}(r, z) = 0 \quad \text{when } r = 0. \tag{2.8}$$

Additionally, the functions of all stresses must be smooth inside of specimen and infinite times differentiable, which allows to state the symmetry of other values. For example the derivatives of u_z and u_r with respect to the z coordinate are even and odd, respectively.

2.1.2 Equilibrium Equations

Taking into account that the cylindrical variables are considered in Euler coordinates, the coordinate system r-z-θ does not change during deformation.

This means that the equilibrium equations are linear and have in cylindrical coordinates the following form:

$$\frac{\partial \sigma_r}{\partial r} + \frac{1}{r}\frac{\partial \tau_{\theta r}}{\partial \theta} + \frac{\partial \tau_{zr}}{\partial z} + \frac{1}{r}(\sigma_r - \sigma_\theta) + f_r = 0, \tag{2.9}$$

$$\frac{\partial \tau_{r\theta}}{\partial r} + \frac{1}{r}\frac{\partial \sigma_\theta}{\partial \theta} + \frac{\partial \tau_{z\theta}}{\partial z} + \frac{2}{r}\tau_{r\theta} + f_\theta = 0, \tag{2.10}$$

$$\frac{\partial \tau_{rz}}{\partial r} + \frac{1}{r}\frac{\partial \tau_{\theta z}}{\partial \theta} + \frac{\partial \sigma_z}{\partial z} + \frac{1}{r}\tau_{rz} + f_z = 0. \tag{2.11}$$

The body forces f_r, f_θ are equal to zero because the specimen is placed vertical during the experiment. In addition, it is assumed that the force f_z is also equal to zero because the stresses created by the gravity force are very small in comparison with the stresses caused by the specimen tension.

In Eqs. 2.9–2.11, the inertial term does not occur because the tensile test is carried out at a very slow strain rate of $\dot{\varepsilon} = 0.00025\ \mathrm{s}^{-1}$ [14]. As a result, the strain rate and the acceleration do not influence the results of the measurement.

Taking advantage of the axial symmetry, Eqs. 2.9–2.11 reduce to two equilibrium equations:

$$\frac{\partial \sigma_r}{\partial r} + \frac{\partial \tau_{zr}}{\partial z} + \frac{1}{r}(\sigma_r - \sigma_\theta) = 0, \tag{2.12}$$

$$\frac{\partial \tau_{rz}}{\partial r} + \frac{\partial \sigma_z}{\partial z} + \frac{1}{r}\tau_{rz} = 0. \tag{2.13}$$

Let us consider these equations in the minimum section plane, i.e. $z = 0$. Taking into account relation (2.7), the last term in Eq. 2.13 equals to zero. Taking into consideration the fact that the stress functions must be smooth and infinite times differentiable, we have:

$$\frac{\partial \tau_{rz}}{\partial r} = 0, \quad 0 \leq r \leq a, \quad z = 0. \tag{2.14}$$

Let us notice that the function σ_z is even with respect to the z variable and its derivative is odd. Therefore, we obtain:

$$\frac{\partial \sigma_z}{\partial z} = 0, \quad 0 \leq r \leq a, \quad z = 0. \tag{2.15}$$

Taking into account the above considerations in the minimum section plane of the sample ($z = 0$), only one equilibrium equation, cf. Eq. 2.12, remains to be satisfied.

2.1.3 Yield Conditions

The yield condition determines the stress state at which the material transforms in the plastic state. The value of the yield stress depends on many parameters such as the strain state, the strain rate, the strain history, the temperature and other. However, in the case of homogeneous and isotropic materials at constant temperature, it is often assumed that k depends only on the strain intensity given in the form [15]:

$$k = k(\varepsilon_{int}, \dot{\varepsilon}_{int}). \tag{2.16}$$

In this consideration it is assumed according to the standard [14] that the tensile test is conducted at a small constant strain rate. Therefore, the model of hardening and ideal plastic materials is obtained, respectively:

$$k = k(\varepsilon_{int}), \quad k = \text{const.} \tag{2.17}$$

At this stage, we will define the strain intensity ε_{int} by means of the second invariant of the strain deviator $J_2(D_\varepsilon)$ as

$$\varepsilon_{int} = \frac{2}{\sqrt{3}} \sqrt{J_2(D_\varepsilon)}, \tag{2.18}$$

where

$$J_2(D_\varepsilon) = \frac{1}{6}[(\varepsilon_r - \varepsilon_\theta)^2 + (\varepsilon_z - \varepsilon_\theta)^2 + (\varepsilon_z - \varepsilon_r)^2]. \tag{2.19}$$

It should be noted that the definition of the strain intensity includes sometimes some other coefficient in front of the root (e.g. 1 instead of $2\sqrt{3}$ as given in [4]). On the base of the analysis made in Sect. 2.1.1, zero terms in relation (2.19) are not considered because of symmetry.

Among the huge variety of yield conditions, two basic equations have a wide application in the practical calculations of plastic forming processes: namely, the condition of the constant value of shearing stress intensity and the condition of the maximal shearing stress [15] which will be discussed in the following.

The condition of the shearing stress intensity is called the Huber–Mises condition and takes in cylindrical coordinates the following form:

$$(\sigma_z - \sigma_r)^2 + (\sigma_r - \sigma_\theta)^2 + (\sigma_\theta - \sigma_z)^2 + 6(\tau_{zr}^2 + \tau_{r\theta}^2 + \tau_{\theta z}^2) = 2k_{HM}^2, \tag{2.20}$$

where k_{HM} is the yield point determined from the tensile test.

However, after taking into account the rotational symmetry with respect to the z axis of the tension specimen (2.6) as well as the sample symmetry in regard to the plane $z = 0$ (2.7), condition (2.20) simplifies to the following form:

$$(\sigma_z - \sigma_r)^2 + (\sigma_r - \sigma_\theta)^2 + (\sigma_\theta - \sigma_z)^2 = 2k_{HM}^2. \tag{2.21}$$

In the principal stress space, this expression is represented by a yield surface in the form of an infinite long cylinder with its principal axis equal to the space diagonal of the positive axes [16].

The yield condition of the maximal shearing stress is also called the Tresca yield condition. In a cylindrical coordinate system it takes the form:

$$\max\{|\sigma_r - \sigma_\theta|, \quad |\sigma_\theta - \sigma_z|, \quad |\sigma_z - \sigma_r|\} = k_T, \tag{2.22}$$

where k_T is the yield point determined from the tensile test.

It is worth noticing that the values of k_{HM} and k_T are not always the same. Their values depend on the kind of experimental test which is carried out for their determination. In the case of tensile test these values are the same, but in the case of shear test they differ more or less by 10% [16].

The yield surface in the principal stress space is in this case an infinite long prism with the cross section of a regular hexagon (see Fig. 2.3 in the Sect. 2.3.2).

2.1.4 Basic Relationships

In Euler coordinates the strain tensor is nonlinear [4] and possesses the following form:

$$\varepsilon_{ij} = \frac{1}{2}\left(\frac{\partial u_i}{\partial x_j} + \frac{\partial u_j}{\partial x_i} - \frac{\partial u_k}{\partial x_i}\frac{\partial u_k}{\partial x_j}\right). \tag{2.23}$$

However, in the case of small displacement gradients, the second (mixed) derivative in Eq. 2.23 can be omitted and the commonly applied small strain tensor is obtained as:

$$\varepsilon_r = \frac{\partial u_r}{\partial r}, \quad \varepsilon_\theta = \frac{u_r}{r}, \quad \varepsilon_z = \frac{\partial u_z}{\partial z}, \quad \varepsilon_{rz} = \frac{1}{2}\left(\frac{\partial u_r}{\partial z} + \frac{\partial u_z}{\partial r}\right). \tag{2.24}$$

The remaining components of the strain tensor are equal to zero in accordance with the symmetry:

$$\varepsilon_{r\theta} = \frac{1}{r}\frac{\partial u_r}{\partial \theta} + \frac{\partial u_\theta}{\partial r} - \frac{u_\theta}{r} \equiv 0, \quad \varepsilon_{z\theta} = \frac{\partial u_\theta}{\partial z} + \frac{1}{r}\frac{\partial u_z}{\partial \theta} \equiv 0. \tag{2.25}$$

The constitutive equation in the deformation theory of plasticity is in general given by

$$D_\varepsilon = \Psi D_\sigma, \tag{2.26}$$

where D_ε and D_σ are the deviators of the strain and stress tensors, and Ψ is the known monotonic function $\Psi = \Psi(J_2(\varepsilon))$ (see for example [8]). This function is determined in the plastic region under the assumption of the Huber–Mises or Tresca yield condition.

Taking advantage of the relation which links a tensor with its isotropic as well as deviatoric components, the separated components can be obtained from Eq. 2.26 as

$$\varepsilon_r - \varepsilon^0 = \Psi(\sigma_r - \sigma^0), \quad \varepsilon_z - \varepsilon^0 = \Psi(\sigma_z - \sigma^0), \quad \varepsilon_\theta - \varepsilon^0 = \Psi(\sigma_\theta - \sigma^0),$$
$$\varepsilon_{rz} = \Psi\tau_{rz}, \tag{2.27}$$

where

$$\sigma^0 = \frac{1}{3}I_1(\sigma), \quad \varepsilon^0 = \frac{1}{3}I_1(\varepsilon). \tag{2.28}$$

Function Ψ is determined in a few stages, as described in detail in book [8]. For example, for the Huber–Mises yield condition its form is represented as follows [8]:

$$\Psi_{HM} = \sqrt{3J_2(\varepsilon)/4k_{HM}^2\{J_2(\varepsilon)\}}, \tag{2.29}$$

which the relation between the first invariants in the deformation theory of plasticity has the form:

$$I_1(\varepsilon) = 3KI_1(\sigma), \tag{2.30}$$

where $I_1(\varepsilon)$ and $I_1(\sigma)$ are the first invariants of the strain and stress tensors, K is the elastic bulk modulus known from the theory of elasticity [11]:

$$K = \frac{3\lambda + 2\mu}{3} = \frac{E}{3(1 - 2v)}, \tag{2.31}$$

where λ and μ are Lamé's elasticity coefficients and v is Poisson's ratio.

2.2 Problem Description based on Plastic Flow Theory

Everything which was presented in Sects. 2.1.1–2.1.3 and related to the displacement vector and the stress tensors holds the same way in plastic flow theory. Additionally, we consider the velocity vector and the strain rate tensor which depends in the considered case only on the cylindrical coordinates r as well as z and do not depend on the θ variable:

$$\bar{\mathbf{v}} = \bar{\mathbf{v}}(r, z), \tag{2.32}$$

$$\underline{\dot{\varepsilon}} = \underline{\dot{\varepsilon}}(r, z). \tag{2.33}$$

First of all, the sample behaviour is only considered in the surrounding of the neck. In this region the elastic strain can be omitted and the derivation is essentially simplified, which leads to the classical plastic flow theory together with the flow rule:

$$D_{\dot\varepsilon} = \lambda \frac{\partial \Phi}{\partial \sigma_{ij}}, \tag{2.34}$$

where λ is the factor of proportionality (sometimes called the plastic multiplier) determined from the yield condition for each stage of deformation and $\Phi(\sigma_{ij})$ is the plastic potential. The relationship

$$I_1(\dot\varepsilon) = 0, \tag{2.35}$$

states that the material is incompressible and the yield condition can be written as

$$f(\sigma_{ij}) = k(\varepsilon_{\text{int}}). \tag{2.36}$$

In the case $\Phi = f$, we obtain the so-called associated flow rule [11]. If the function $\Phi(\sigma_{ij})$ is not smooth in all ranges, so that Eq. 2.34 cannot be applied in these points, it must be proceeded in such a manner as it will be described for the case of the Tresca yield condition (see Sect. 2.3.2).

We assume in accordance with the theory [17] that the strain rate tensor is linear:

$$\dot\varepsilon_r = \frac{\partial v_r}{\partial r}, \quad \dot\varepsilon_\theta = \frac{v_r}{r}, \quad \dot\varepsilon_z = \frac{\partial v_z}{\partial z}, \quad \dot\varepsilon_{rz} = \frac{1}{2}\left(\frac{\partial v_r}{\partial z} + \frac{\partial v_z}{\partial r}\right), \quad \dot\varepsilon_{\theta z} = \dot\varepsilon_{\theta r} = 0, \tag{2.37}$$

where the strain tensor is calculated in a different way compared to the deformation theory, since the integral from the strain rate tensor has the form:

$$\varepsilon_{ij} = \int_{t_0}^{t} \dot\varepsilon_{ij}(t, x)\, dt. \tag{2.38}$$

2.3 Formulae Derivation based on Bridgman Approach

2.3.1 Derivation of the Basic Relationships in Deformation Theory

In order to determine the stress distribution in the neck of the tensile sample (Fig. 2.1), Bridgman assumed on the base of experiments [1, 2] that the radial displacement varies linearly in the minimal section plane as

$$u_r(r, z) \cong A(z)r, \quad |z| \le \vartheta. \tag{2.39}$$

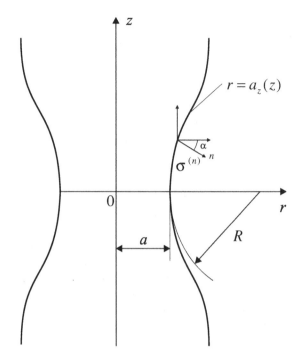

Fig. 2.1 The geometry of deformed sample

Therefore, calculating the radial and circumferential strains from Eq. 2.24 we obtain:

$$\varepsilon_r = A(z), \quad \varepsilon_\theta = A(z). \tag{2.40}$$

These relations enable to formulate the simplification, being a result of the assumption given in Eq. 2.39.

The First Simplifying Assumption

In the minimal section plane the circumferential strain is equal to the radial strain

$$\varepsilon_\theta = \varepsilon_r. \tag{2.41}$$

Assumption (2.41), together with the deformation theory of plasticity, allows to conclude to a certain relation regarding the stresses. Namely, from Eqs. 2.27 to 2.41 it follows that:

$$\sigma_r = \sigma_\theta. \tag{2.42}$$

Relation (2.42) allows to transform the equilibrium equation (2.12) to obtain:

$$\frac{\partial \sigma_r}{\partial r} + \frac{\partial \tau_{zr}}{\partial z} = 0 \quad \text{when } z = 0. \tag{2.43}$$

Let us consider separately the Huber–Mises and Tresca yield conditions.

Taking advantage of relation (2.42) in the Huber–Mises yield condition (2.21), we get:

$$(\sigma_z - \sigma_r)^2 + (\sigma_r - \sigma_z)^2 = 2k_{HM}^2, \tag{2.44}$$

and after simplification we obtain:

$$(\sigma_z - \sigma_r)^2 = k_{HM}^2. \tag{2.45}$$

Extracting the root from both sides of expression (2.45), we obtain:

$$|\sigma_z - \sigma_r| = k_{HM}, \tag{2.46}$$

while taking into account the relation (2.42) from the Tresca yield condition (2.22), we have:

$$\max\{0, |\sigma_r - \sigma_z|, \quad |\sigma_z - \sigma_r|\} = k_T. \tag{2.47}$$

Therefore, from Eq. 2.47 it follows that:

$$|\sigma_z - \sigma_r| = k_T \tag{2.48}$$

The following conclusion can be drawn from relations (2.46) and (2.48): both yield conditions under consideration of the additional first simplifying assumption and the derived relation (2.42) can be written in the same form.

As it was mentioned in Sect. 2.1.3 the constant values k_{HM} and k_T determined from the tensile test are the same. Therefore, in the further considerations the constants will be no longer distinguished: the notation of k will be used in the following for the yield stress. It was underlined in Sect. 2.1.3 that $k = k(\varepsilon_{int})$ and, generally speaking, it is not constant because $\varepsilon_{int} = \varepsilon_{int}(r, z)$.

Finally, the Huber–Mises (2.46) and Tresca (2.48) yield conditions can be written in the form:

$$\sigma_z - \sigma_r = k. \tag{2.49}$$

Taking advantage of the fact that for a tensile test in the direction of the z axis it is natural that $\sigma_z \geq \sigma_r$, condition (2.49) is also satisfied in the minimum section plane, i.e.:

$$\sigma_z = k + \sigma_r \quad \text{for } z = 0, \tag{2.50}$$

which can be differentiated with respect to the r variable to obtain:

$$\frac{\partial \sigma_z}{\partial r} = \frac{\partial k}{\partial r} + \frac{\partial \sigma_r}{\partial r} \quad \text{for } z = 0. \tag{2.51}$$

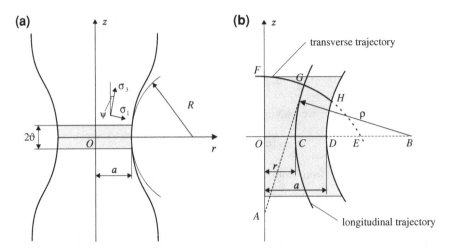

Fig. 2.2 Neck geometry in a tension sample (**a**) principal stress trajectories in the meridian surface (**b**)

The Second Simplifying Assumption

Following Bridgman, we assume that in a certain surrounding of the neck the value of the yield stress k is constant (this region is shown in the Fig. 2.2a):

$$k(r, z) = k_* \quad \text{for } 0 < r < a, \ |z| \le \vartheta. \tag{2.52}$$

It follows that

$$\left. \frac{\partial k}{\partial r} \right|_{z=0} = \left. \frac{\partial k}{\partial z} \right|_{z=0} = 0. \tag{2.53}$$

It needs to be noticed that the value of k_* changes in general during the deformation (in the case of material hardening). However, the assumption $\left. \frac{\partial k}{\partial z} \right|_{z=0} = 0$ is always satisfied if the function $k(r, z)$ is differentiable, which follows from the symmetry assumption.

Returning to Eq. 2.51 and taking into account the relationship (2.53), one can obtain that:

$$\frac{\partial \sigma_z}{\partial r} = \frac{\partial \sigma_r}{\partial r}. \tag{2.54}$$

Taking into consideration the relation (2.54), Eq. 2.43 takes the following form:

$$\frac{\partial \sigma_z}{\partial r} + \frac{\partial \tau_{zr}}{\partial z} = 0 \quad \text{when } z = 0. \tag{2.55}$$

Let us consider the principal stress trajectory σ_3 in the meridian surface defined as a plane section of a revolution surface containing the axis of revolution. Let us denote the slope angle of the trajectory tangent to the z axis by ψ (Fig. 2.2).

Subsequently in accordance with formulae given for instance in [8, 10] for the stress in the surface perpendicular to one of the principal surfaces, where the normal creates the angle ψ with the principal axis, one can obtain:

$$\sigma_z = \frac{\sigma_3 + \sigma_1}{2} + \frac{\sigma_3 - \sigma_1}{2} \cos 2\psi, \tag{2.56}$$

$$\sigma_r = \frac{\sigma_3 + \sigma_1}{2} - \frac{\sigma_3 - \sigma_1}{2} \cos 2\psi, \tag{2.57}$$

$$\tau_{zr} = \frac{\sigma_3 - \sigma_1}{2} \sin 2\psi. \tag{2.58}$$

Whereas, the component of the circumferential stress σ_θ is determined from relation (2.42).

Let us notice following Bridgman that the longitudinal trajectory is perpendicular to the minimum section. From thus it follows that:

$$\psi(r, 0) = 0. \tag{2.59}$$

The Third Simplifying Assumption

Now we assume that in a small surrounding of the minimum section plane $z = 0$ (see Fig. 2.2) the angle ψ is small enough, i.e.

$$\psi(r, z) \ll 1, \quad 0 < r < a, \quad |z| < \tilde{\vartheta}. \tag{2.60}$$

Taking advantage of the third simplifying assumption, we obtain that $\cos 2\psi \approx 1$ and $\sin 2\psi \approx 2\psi$. Thus, from Eqs. 2.56–2.58 and the yield condition (2.49) we have:

$$\sigma_r \approx \sigma_1, \quad \sigma_z \approx \sigma_3, \quad \tau_{rz} \approx (\sigma_3 - \sigma_1)\psi \approx k\psi, \tag{2.61}$$

where σ_1 and σ_3 are the principal stresses in the meridian surface.

It is worth noticing that condition (2.15) together with assumption $(2.61)_2$ means that the longitudinal trajectory is perpendicular to the minimum section plane. Therefore, the third simplifying assumption is true.

Let us consider the second component of Eq. 2.55 in the form:

$$\left(\frac{\partial \tau_{rz}}{\partial z}\right)_{z=0} = \left(\frac{\partial(k\psi)}{\partial z}\right)_{z=0} = k\left(\frac{\partial \psi}{\partial z}\right)_{z=0} + \underbrace{\psi \overbrace{\frac{\partial k}{\partial z}\Big|_{z=0}}^{0}}_{0} = k\left(\frac{\partial \psi}{\partial z}\right)_{z=0}, \tag{2.62}$$

where relation (2.53), which follows from the second simplifying assumption, has been utilised.

Since the angle ψ is small enough (see the third simplifying assumption, which indeed is not already the assumption) we obtain:

$$\psi(r,z) \approx \tan\psi(r,z) = f'_C(z),$$ (2.63)

where $r = f_C(z)$ is the appropriate longitudinal trajectory passing through point C on the OB axis (Fig. 2.2b). Calculating the derivative from Eq. 2.63 we obtain:

$$\left(\frac{\partial\psi}{\partial z}\right) = f''(z).$$ (2.64)

The curvature of the principal stress trajectory σ_3 is given by the formula (see for example [12]):

$$\frac{1}{\rho} = \frac{|f''(z)|}{(1+f'(z)^2)^{\frac{3}{2}}},$$ (2.65)

which can be transformed to obtain:

$$f''(z) = \frac{(1+f'(z)^2)^{\frac{3}{2}}}{\rho}.$$ (2.66)

Inserting Eqs. 2.64 and 2.66 into 2.62 with reference 2.63 we obtain:

$$k\left(\frac{\partial\psi}{\partial z}\right)_{z=0} = k\left[\frac{(1+f'(z)^2)^{\frac{3}{2}}}{\rho}\right]_{z=0} = k\left[\frac{(1+\psi^2(r,z))^{\frac{3}{2}}}{\rho}\right]_{z=0}.$$ (2.67)

In the plane $z = 0$ the angle is zero, i.e. $\psi = 0$ (in accordance with (2.59)), therefore Eq. 2.67 can be simplified to the form:

$$k\left(\frac{\partial\psi}{\partial z}\right)_{z=0} = \frac{k}{\rho}.$$ (2.68)

Finally, Eq. 2.62 can be written in the following form:

$$\left.\frac{\partial\tau_{rz}}{\partial z}\right|_{z=0} = \frac{k}{\rho}, \quad 0 \leq r \leq a.$$ (2.69)

Returning to Eq. 2.55 and inserting it into relation (2.69), it reduces finally to the form:

$$\frac{\partial\sigma_z}{\partial r} + \frac{k}{\rho} = 0 \quad \text{when } z = 0,\ 0 < r < a.$$ (2.70)

In order to complete the solution, the relation $\rho = \rho(r)$ needs to be determined.

The Fourth Simplifying Assumption

Bridgman assumed that in the surrounding of the minimum section, the transverse trajectories of the principal stress are arcs which are orthogonal to the longitudinal trajectories.

Let *FH* be such a transverse trajectory intersecting the minimum section *OD* externally at point *E* and having its curvature centre at point *A* (see Fig. 2.2b). If *B* is the curvature centre of the longitudinal trajectory, one can conclude that:

$$\rho^2 = BG^2 = AB^2 - AE^2 = OB^2 - OE^2 = (r + \rho)^2 - OE^2, \qquad (2.71)$$

where *r* is the radial distance of point *C* from the longitudinal axis.

Formula (2.71) is true for any point *G* on the circle *FGH*, including the point *H*. It allows to create the relation: $R^2 = (a + R)^2 - OE^2$ which takes together with Eq. 2.71 the following form:

$$r^2 + 2r\rho = a^2 + 2aR = \text{const}(= OE^2), \qquad (2.72)$$

where *a* is the sample radius in the minimum section plane. Subsequently, the relationship $\rho = \rho(r)$ is obtained as

$$\rho = \frac{a^2 + 2aR - r^2}{2r}. \qquad (2.73)$$

Inserting relation (2.73) into (2.70) we obtain:

$$\frac{\partial \sigma_z}{\partial r} + \frac{2rk}{a^2 + 2aR - r^2} = 0. \qquad (2.74)$$

The boundary condition on the neck contour has the form:

$$\sigma_r = 0, \quad \tau_{rz} = 0 \quad \text{at } z = 0, \ r = a, \qquad (2.75)$$

which by means of expression (2.49) can be represented in the form:

$$\sigma_z = k, \quad \text{at } z = 0 \text{ and } r = a. \qquad (2.76)$$

Separating variables in Eq. 2.74, the expression can be written as

$$d\sigma_z = -\frac{2rk}{a^2 + 2aR - r^2} \, dr. \qquad (2.77)$$

Last equation can be integrated on both sides under utilisation of the boundary condition (2.76) to finally obtain the formula for the normalised axial stress in the minimum section plane in the following form:

$$\frac{\sigma_z}{k} = 1 + \ln\left(1 + \frac{a^2 - r^2}{2aR}\right). \qquad (2.78)$$

The average axial stress $\bar{\sigma}_z$ in the minimum section is given by the formula [7]:

$$\bar{\sigma}_z = \frac{2}{a^2} \int_0^a \sigma_z r \, dr. \tag{2.79}$$

Inserting relation (2.78) into (2.79), we obtain:

$$\frac{\bar{\sigma}_z}{k} = \frac{2}{a^2} \int_0^a \left(1 + \ln\left(1 + \frac{a^2 - r^2}{2aR} \right) \right) r \, dr. \tag{2.80}$$

As a result of integration by parts of Eq. 2.80, we finally get the formula for the average normalised axial stress in the form:

$$\frac{\bar{\sigma}_z}{k} = \left(1 + \frac{2R}{a} \right) \ln\left(1 + \frac{a}{2R} \right). \tag{2.81}$$

In order to calculate the average normalised axial stress based on measurements from experimental tests, the following formula should be utilised:

$$F = \pi a^2 \bar{\sigma}_z. \tag{2.82}$$

In addition to the formulae for the normalised axial stress (2.78) and the average normalised axial stress (2.81) in the minimum section, it is also worth deriving relations for the normalised radial and circumferential stresses. For that purpose relation (2.54) with boundary condition (2.75) will be applied in Eq. 2.74 to obtain:

$$\frac{1}{k} \int_{\sigma_r}^0 d\sigma_r = -2 \int_r^a \frac{r}{a^2 + 2aR - r^2} dr. \tag{2.83}$$

After integrating, we finally obtain the formula for the normalised radial stress in the minimum section plane in the following form:

$$\frac{\sigma_r}{k} = \ln\left(1 + \frac{a^2 - r^2}{2aR} \right). \tag{2.84}$$

Taking advantage of relation (2.42) on the equality of the radial and circumferential stresses, we obtain the formula for the normalised circumferential stress in the same form as for the normalised radial stress (see formula (2.84)).

Let us notice that other authors [7, 8] generalised the Bridgman consideration on the plastic flow law with Huber–Mises and Tresca yield conditions, which is discussed in the following section.

2.3.2 Derivation of the Basic Relations in Plastic Flow Theory

As a result of the application of the simplifying assumptions and the deformation theory of plasticity, Bridgman obtained the two Eqs. 2.42 and 2.49. It should be proven that these relationships are also true in scope of the plastic flow theory. Let us consider them for each yield condition separately.

As it was mentioned earlier, it follows from experiments [1, 2] that the displacement as well as the radial velocity change approximately linear in the minimum section plane of the specimen:

$$v_r(r,z) \cong A(z)r \quad \text{for } |z| \leq \vartheta. \tag{2.85}$$

Therefore, calculating the radial and circumferential strain rates from Eq. 2.37 gives:

$$\dot{\varepsilon}_r = A(z), \quad \dot{\varepsilon}_\theta = A(z). \tag{2.86}$$

These relations allow to formulate the following simplification:

The First Simplifying Assumption

The circumferential strain rate is equal to the radial strain rate in the minimal section plane:

$$\dot{\varepsilon}_\theta = \dot{\varepsilon}_r. \tag{2.87}$$

Writing the relation between strain rates and the stresses in the case of Huber–Mises and the associated flow law, one obtains:

$$\dot{\varepsilon}_\theta = \lambda(\sigma_\theta - \sigma^0), \quad \dot{\varepsilon}_r = \lambda(\sigma_r - \sigma^0), \quad \dot{\varepsilon}_z = \lambda(\sigma_z - \sigma^0), \quad \sigma^0 = \frac{1}{3}I_1(\sigma). \tag{2.88}$$

Opposed to the assumptions of the deformation theory of plasticity (2.27), it is here assumed that the material is incompressible (2.35) which means that $\dot{\varepsilon}^0 = 0$.

Therefore, it follows from Eqs. 2.87 and 2.88 that the condition (2.42) is satisfied.

Having proven assumption (2.42) and the Huber–Mises yield condition in Eqs. 2.44–2.46, it is proven that $|\sigma_z - \sigma_r| = k_{HM}$.

Let us consider now the Tresca yield condition. In this case, the yield surface in the principal stress space is an infinitely long prism with regular hexagon as cross section. Figure 2.3 presents a view of this prism in the principal stress space. In the case when the stress state is represented by a point placed on the smooth part of this surface (for instance points A, C in the Fig. 2.3), the relations (2.88) are satisfied and we obtain the Tresca yield condition according to Eq. 2.48.

However, if the stress state corresponds to a corner or edge on the prism, the strain rate vector can take any direction situated between the orthogonal directions of the surfaces adjacent to the irregular point (e.g. point B in the Fig. 2.3). In this

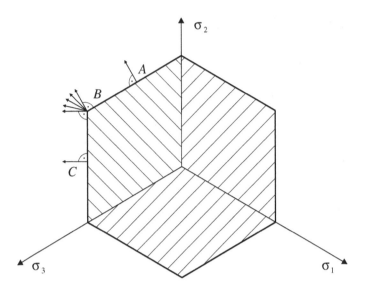

Fig. 2.3 Tresca yield surface in the principal stress space

case, we cannot utilise relation (2.88) and we must carry out an analysis as described in the following.

Let us notice that the stress tensor in the axisymmetric case has the form:

$$\underline{\sigma} = \begin{bmatrix} \sigma_r & 0 & \tau_{rz} \\ 0 & \sigma_\theta & 0 \\ \tau_{rz} & 0 & \sigma_z \end{bmatrix}, \tag{2.89}$$

from which follows that $\sigma_\theta = \sigma_2$. However, the remaining principal stresses σ_3 and σ_1 are determined from the equation:

$$\begin{bmatrix} \sigma_r - \sigma & \tau_{rz} \\ \tau_{rz} & \sigma_z - \sigma \end{bmatrix} = 0, \tag{2.90}$$

which can be transformed to obtain:

$$\sigma^2 - (\sigma_r + \sigma_z)\sigma + \sigma_r\sigma_z - \tau_{rz}^2 = 0. \tag{2.91}$$

Determining the roots of Eq. 2.91 gives:

$$\sigma_{1,3} = \frac{\sigma_r + \sigma_z \pm \sqrt{(\sigma_r + \sigma_z)^2 - 4\sigma_r\sigma_z + 4\tau_{rz}^2}}{2}. \tag{2.92}$$

The Tresca yield condition formulated in principal stresses under the assumption $\sigma_1 < \sigma_\theta < \sigma_3$ takes the following form:

$$\sigma_3 - \sigma_1 = k_T, \tag{2.93}$$

which after replacement of the stress components in cylindrical coordinates takes the form:

$$(\sigma_z - \sigma_r)^2 + 4\tau_{rz}^2 = k^2 \quad \text{or} f(\sigma_{ij}) = \sqrt{(\sigma_z - \sigma_r)^2 + 4\tau_{rz}^2}. \tag{2.94}$$

In order to determine the plastic flow on the yield surface, the following relation must be satisfied:

$$\dot{\varepsilon}_{ij} = \lambda \frac{\partial \Phi}{\partial \sigma_{ij}}, \tag{2.95}$$

where Φ is the plastic potential. Let us notice that this potential is smooth with respect to σ_r, σ_z and τ_{rz}.

Calculating $\dot{\varepsilon}_r, \dot{\varepsilon}_z$, and $\dot{\varepsilon}_{rz}$ from (2.94) using (2.95) in the case of the associated flow law $\Phi = f$, we obtain:

$$\dot{\varepsilon}_r = -\tilde{\lambda}(\sigma_z - \sigma_r), \quad \dot{\varepsilon}_z = \tilde{\lambda}(\sigma_z - \sigma_r), \quad \dot{\varepsilon}_{rz} = 4\tilde{\lambda}\tau_{rz}, \tag{2.96}$$

where

$$\tilde{\lambda} = \lambda \frac{1}{\sqrt{(\sigma_z - \sigma_r)^2 + 4\tau_{rz}^2}}. \tag{2.97}$$

From Eqs. 2.96_1 and 2.96_2, it is easy to conclude that:

$$\dot{\varepsilon}_r + \dot{\varepsilon}_z = 0. \tag{2.98}$$

Unfortunately, Eq. 2.95 cannot be applied for the component $\dot{\varepsilon}_\theta$. However, from the material incompressibility assumption, according to Eq. 2.35, it follows that:

$$\dot{\varepsilon}_\theta = 0. \tag{2.99}$$

Note that the last expression has been obtained without utilisation of Eq. 2.95.

Let us consider the principal stresses in the minimum section plane, from Eq. 2.92 and taking advantage of Eq. 2.7 we obtain:

$$\sigma_{1,3} = \frac{\sigma_r + \sigma_z \pm \sqrt{(\sigma_r - \sigma_z)^2}}{2}, \quad \sigma_1 = \sigma_r, \quad \sigma_3 = \sigma_z. \tag{2.100}$$

We took into account the stress analysis in this cross section, from which it follows that $\sigma_3 \geq \sigma_1$ holds. To sum up, in the minimum section plane the relationship $\sigma_3 - \sigma_1 = |\sigma_z - \sigma_r| = k_T$ holds (see also (2.42)).

Therefore, condition (2.48) has been proven in the case of Tresca for a point laying on an edge, but the proofs were based on other considerations. Simultaneously, the condition (2.49) is also in both cases proven.

The derivation of the basic relations according to Bridgman in the further part proceeds analogous to Sect. 2.3.1, where the deformation theory of plasticity has been applied.

2.4 Formulae Derivation in Davidenkov and Spiridonova Approach

The derivation of the stress formulae proceeds in the same manner as in the Bridgman approach as far as Eq. 2.70 is concerned. However, the relation for $\rho = \rho(r)$ is presented in another way. Namely, Davidenkov and Spiridonova assumed that ρ is inverse proportional to r and it is expressed by the formula:

$$\rho = \frac{Ra}{r}. \tag{2.101}$$

It is easy to notice that for such an assumption the boundary conditions are satisfied as in the case of the Bridgman approach (2.73):

$$\lim_{r \to 0} \rho(r) = \infty, \quad \lim_{r \to a} \rho(r) = R. \tag{2.102}$$

Inserting Eq. 2.101 into 2.70 we obtain:

$$\frac{\partial \sigma_z}{\partial r} + \frac{rk}{Ra} = 0, \tag{2.103}$$

or

$$d\sigma_z = -\frac{rk}{Ra} dr, \tag{2.104}$$

for the boundary conditions given in Eq. 2.76.

Equation 2.104 can be easily integrated to finally obtain:

$$\frac{\sigma_z}{k} = 1 + \frac{a^2 - r^2}{2Ra}. \tag{2.105}$$

Utilising the formula for the average axial stress $\bar{\sigma}_z$ in the minimum section plane as given in Eq. 2.79, one can obtain:

$$\frac{\bar{\sigma}_z}{k} = \frac{2}{a^2} \int_0^a \left(1 + \frac{a^2 - r^2}{2aR}\right) r \, dr. \tag{2.106}$$

After integrating Eq. 2.106, we finally get the formula for the average normalised axial stress in the form

$$\frac{\bar{\sigma}_z}{k} = 1 + \frac{a}{4R}. \tag{2.107}$$

In order to derive the relation for the normalised radial stress in the minimum section plane, we can apply formula (2.54) with boundary condition (2.75) in Eq. 2.103 to obtain:

$$\frac{1}{k}\int\limits_{\sigma_r}^{0} d\sigma_r = -\int\limits_{r}^{a} \frac{r}{Ra}dr. \tag{2.108}$$

Finally, after integrating, the formula for the normalised radial stress has the following form:

$$\frac{\sigma_r}{k} = \frac{a^2 - r^2}{2Ra}, \tag{2.109}$$

and after utilising the relation (2.42), the formula for the normalised circumferential stress has the same form as for the normalised radial stress according to Eq. 2.109.

2.5 Formulae Derivation in Siebel Approach

The derivation of the relation for the stress distribution in the minimal section plane according to Siebel [13] proceeds analogous to the Bridgman approach as far as Eq. 2.70 is concerned. However, he assumed a more general form for the curvature radius relation of the longitudinal stress trajectory compared to the Davidenkov and Spiridonova approach given in (2.101). Namely, on the base of the trajectory behaviour in the surrounding of the minimal section plane (see Eq. 2.102), Siebel assumed that:

$$\rho = R\left(\frac{a}{r}\right)^n, \tag{2.110}$$

which is in accordance with Eq. 2.102. Inserting Eq. 2.110 into 2.70, we obtain:

$$\frac{\partial \sigma_z}{\partial r} + \frac{k}{R}\left(\frac{r}{a}\right)^n = 0, \tag{2.111}$$

or

$$d\sigma_z = -\frac{k}{R}\left(\frac{r}{a}\right)^n dr, \tag{2.112}$$

for the boundary conditions given in Eq. 2.76. Equation 2.112 can be easily integrated to finally obtain the expression:

$$\frac{\sigma_z}{k} = 1 + \frac{a^{n+1} - r^{n+1}}{(n+1)Ra^n}.$$ (2.113)

Taking advantage of formula (2.79) for the average normalised axial stress in the minimal section plane, one can obtain:

$$\frac{\bar{\sigma}_z}{k} = \frac{2}{a^2} \int\limits_0^a \left(1 + \frac{a^{n+1} - r^{n+1}}{(n+1)Ra^n} \right) r \, dr.$$ (2.114)

As a result of integration of Eq. 2.114, we finally obtain the formula for the average normalised axial stress in the form:

$$\frac{\bar{\sigma}_z}{k} = 1 + \frac{a}{(n+3)R}.$$ (2.115)

Siebel noticed that at the free sample surface the curvature radius of the principal stress trajectories are linearly linked with the curvature radius of the contour. As a result, the exponent n is here equal to 1. However, the influence of the exponent n on the average axial stress in the narrow cross section is relatively small. Therefore, Siebel assumed in his solution that $n = 0$ in the minimum section plane.

The final form of his solution overlaps with the proposed Eq. 2.107 by Davidenkov and Spiridonova. For this reason, formula (2.107) will be called in the further part of this monograph the Siebel–Davidenkov–Spiridonova formula. Obviously, formulae for normalised axial, radial and circumferential stresses in the Siebel approach will correspond to expressions of the Davidenkov-Spiridonova solutions given in the form of Eqs. 2.105 and 2.109, respectively.

2.6 Formulae Derivation in Szczepiński Approach

Szczepiński wrote in his book [3] that the equilibrium equation of a small volume element near the principal surfaces can be written in the following approximate form:

$$\frac{\sigma_z}{\rho} - \frac{\sigma_\theta - \sigma_r}{r} + \frac{d\sigma_r}{dr} = 0, \quad z = 0.$$ (2.116)

Unfortunately, the authors of this monograph did not succeed to find the original derivation. Thus, the attempt will be made in this subsection to derive this simplified equilibrium equation based on two different approaches.

The First Attempt

In the scope of the first approach, a small element cut with principal longitudinal surfaces bounded on both sides with the surfaces including the curvature radius of the longitudinal stress trajectories creating an angle $d\beta$ is considered. Furthermore, two surfaces in the circumferential direction form an angle $d\theta$ (Fig. 2.4). The coaxiality of the longitudinal trajectories is assumed and constant stresses are acting perpendicular to the given cross section.

In order to write the equilibrium equation of the considered element, it is necessary to determine the formulae for the surface areas of the element, on which the stresses are acting. The sector $ABCD$ for a small value of the angle $d\beta/2$ can be considered as the element section in the minimal section plane (Fig. 2.4b). It can be calculated as the sector between two circles in the form $P_{ABCD} = r\,dr\,d\theta$. However, the surface area of the sector $ABEF$ is estimated by the formula $P_{ABEF} = \rho\,dr\,d\beta$. The remaining surface areas to state the variation of stresses are function of the angle $d\beta$. They are determined in the form of integrals and will be written directly in the equilibrium equation. Therefore, projecting forces acting on the element in direction of the r axis, one obtains:

$$\sigma_3 \sin(d\beta/2)r\,dr\,d\theta + (\sigma_3 + d\sigma_3)\sin(d\beta/2)r\,dr\,d\theta$$

$$- r\,d\theta \int_{-d\beta/2}^{d\beta/2} \sigma_1 \cos\varphi(\rho + dr)\,d\varphi$$

$$+ (r + dr)\,d\theta \int_{-d\beta/2}^{d\beta/2} (\sigma_1 + d\sigma_1)\cos\varphi\rho\,d\varphi$$

$$- \sigma_\theta \sin\frac{d\theta}{2}\rho\,dr\,d\beta - (\sigma_\theta + d\sigma_\theta)\sin\frac{d\theta}{2}\rho\,dr\,d\beta = 0, \quad (2.117)$$

where all stresses in the r direction were taken into account. However, in the remaining directions the equilibrium exists following from the symmetry of the problem. In a small surrounding of the minimum section, the angles $d\beta$, φ and $d\theta$ take small values and we get therefore: $\sin(d\beta/2) \approx d\beta/2$, $\cos\varphi \approx 1$, $\sin(d\theta/2) \approx d\theta/2$ as well as taking into account the fact that $d\sigma_3\,d\theta$ and $d\sigma_\theta\,d\beta$ are small in comparison with the remaining components, Eq. 2.117 can be written in the following form:

$$\sigma_3 r\,dr\,d\beta\,d\theta - r\,d\theta \int_{-d\beta/2}^{d\beta/2} \sigma_1(\rho + dr)\,d\varphi + (r + dr)d\theta \int_{-d\beta/2}^{d\beta/2} (\sigma_1 + d\sigma_1)\rho\,d\varphi$$

$$- \sigma_\theta\rho\,dr\,d\beta\,d\theta = 0,$$

$$(2.118)$$

Fig. 2.4 Stress state in a
small element cut with
principal longitudinal
surfaces (**a**) stress state in the
minimum section plane of the
specimen (**b**)

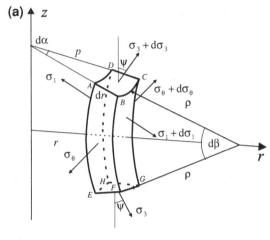

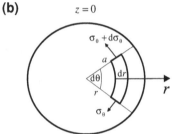

where the expressions under the integrals are constant. Taking into account the
stress constancy, the integrals can be calculated and then dividing relation (2.118)
by the product $d\beta \, d\theta$ we obtain:

$$\sigma_3 r \, dr - r\sigma_1(\rho + dr) + (r + dr)(\sigma_1 + d\sigma_1)\rho - \sigma_\theta \rho \, dr = 0. \qquad (2.119)$$

After simplification and omission of the term $d\sigma_1 \rho \, dr$, since it can be taken as a
small value in comparison with the remaining terms and then dividing by $\rho r \, dr$,
one gets the following expression:

$$\frac{\sigma_3 - \sigma_1}{\rho} - \frac{\sigma_\theta - \sigma_1}{r} + \frac{d\sigma_1}{dr} = 0. \qquad (2.120)$$

Utilising the relations linking the principal with the radial and axial stresses in
the surrounding of the minimal section plane (2.61), we finally obtain the simplified equilibrium equation in the form:

$$\frac{\sigma_z - \sigma_r}{\rho} - \frac{\sigma_\theta - \sigma_r}{r} + \frac{d\sigma_r}{dr} = 0. \qquad (2.121)$$

As it is visible, this relation differs from the expression derived by Szczepiński
(2.116) by the term σ_r/ρ.

The Second Attempt

Taking into account that Szczepiński did not determine exactly how the bounding surfaces of the element should look on the top and bottom, an additional approach has been improved in which it has been considered that the small element cut with principal longitudinal surfaces is bounded on both sides by surfaces of a distance of $\pm$ dz. Analogous to the first approach, the two surfaces should form in the circumferential direction an angle dθ. The new element contour considered in this approach is marked in Fig. 2.5 by a solid line and for comparison the element profile from the earlier approach is indicated by the dashed line together with the stress components. In this approach, the assumption of the coaxiality of the longitudinal trajectories is omitted.

The surface area of the sector $A'BC'D'$ is calculated in an analogous way to the first approach and it equals to $P_{A'BC'D'} = r$ dr dθ. However, the surface area of the sector $A'BE'F$ is determined as the area of a rectangle 2dz dr to which the sector area of the arc $A'E'$ should be added, which is bounded by a line parallel to the z axis. In addition, the sector area of the arc BF, which is bounded by a line parallel to the z axis, must be subtracted. The sum of all these sector areas equals to dz^3 dρ/$2\rho(\rho + d\rho)$. However, this is very small in comparison with the rectangle area and we finally assume that the surface area of the sector is $P_{A'BE'F} = 2$dz dr. The remaining surface areas to state the variation of stresses are functions of dβ as well as dγ and are determined in the form of integrals and will be directly written in the equilibrium equation. Therefore, projecting forces acting on the element in the direction of the r axis, one can obtain:

$$2(\tau_{rz}+\mathrm{d}\tau_{rz})r\mathrm{d}r\,\mathrm{d}\theta - r\,\mathrm{d}\theta \int_{-\mathrm{d}\beta/2}^{\mathrm{d}\beta/2} \sigma_1\cos\varphi\rho\,\mathrm{d}\varphi$$

$$+(r+\mathrm{d}r)(\rho+\mathrm{d}\rho)\,\mathrm{d}\theta \int_{-\mathrm{d}\gamma/2}^{\mathrm{d}\gamma/2} (\sigma_1+\mathrm{d}\sigma_1)\cos\eta\mathrm{d}\eta$$

$$-\sigma_\theta\sin\frac{\mathrm{d}\theta}{2}(2\,\mathrm{d}z\,\mathrm{d}r) - (\sigma_\theta+\mathrm{d}\sigma_\theta)\sin\frac{\mathrm{d}\theta}{2}(2\,\mathrm{d}z\,\mathrm{d}r)=0. \quad (2.122)$$

Utilising the fact that in a surrounding of the minimum section plane, the angles dθ, φ and η take small values, i.e.: $\sin(\mathrm{d}\theta/2) \approx \mathrm{d}\theta/2$, $\cos\varphi \approx 1$ and $\cos\eta \approx 1$, Eq. 2.122 simplifies to the form:

$$2\tau_{rz}r\,\mathrm{d}r\,\mathrm{d}\theta - r\,\mathrm{d}\theta \int_{-\mathrm{d}\beta/2}^{\mathrm{d}\beta/2} \sigma_1\rho\mathrm{d}\varphi + (r+\mathrm{d}r)(\rho+\mathrm{d}\rho)\mathrm{d}\theta \int_{-\mathrm{d}\gamma/2}^{\mathrm{d}\gamma/2} (\sigma_1+\mathrm{d}\sigma_1)\,\mathrm{d}\eta$$

$$- 2\sigma_\theta\,\mathrm{d}\theta\,\mathrm{d}z\,\mathrm{d}r = 0, \quad\quad\quad\quad\quad\quad\quad\quad\quad\quad\quad (2.123)$$

where it was taken into account that d$\tau_{rz}r$ dr dθ and dσ_θ dθ dz dr are infinitesimals and can be omitted. Calculating the integrals and then diving by dθ, we have:

Fig. 2.5 Stress states of small elements based on both approaches (**a**) stress state of the considered element in the second approach (**b**)

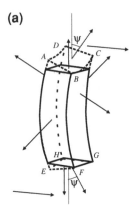

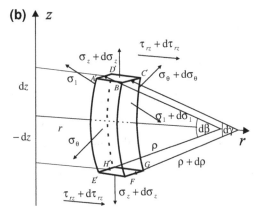

$$2\tau_{rz} r\,dr - \sigma_1 \rho r\,d\beta + (\sigma_1 + d\sigma_1)(r + dr)(\rho + d\rho)d\gamma - 2\sigma_\theta dz\,dr = 0. \quad (2.124)$$

On the base of Fig. 2.5b, it can be easily determined that $d\beta = 2dz/\rho$ and $d\gamma = 2dz/(\rho + d\rho)$, which allow to simplify Eq. 2.124 to the following form:

$$\tau_{rz} r\,dr - \sigma_1 r\,dz + (\sigma_1 + d\sigma_1)(r + dr)dz - \sigma_\theta dz\,dr = 0. \quad (2.125)$$

After simplification and dividing by dz, we have:

$$\frac{\tau_{rz} r\,dr}{dz} + \sigma_1\,dr + d\sigma_1 r + d\sigma_1\,dr - \sigma_\theta\,dr = 0. \quad (2.126)$$

Omitting the small term $d\sigma_1\,dr$ and dividing by $r\,dr$, we finally obtain:

$$\frac{\tau_{rz}}{dz} - \frac{\sigma_\theta - \sigma_1}{r} + \frac{d\sigma_1}{dr} = 0. \quad (2.127)$$

In the surrounding of the minimum section plane, we can apply the formula linking the shearing and principal stresses in the form of Eq. 2.61 for $\psi = d\beta/2$ (see Fig. 2.5). Transforming $\tau_{rz}/dz = (\sigma_3 - \sigma_1)\psi/dz = (\sigma_3 - \sigma_1)d\beta/2dz$ and taking into account that $d\beta = 2dz/\rho$, we have: $\tau_{rz}/dz = (\sigma_3 - \sigma_1)/\rho$. Inserting this relation into the equilibrium equation (2.127), we obtain again formula (2.120) from the first attempt!

In the case of receiving the same solution based on two different approaches, it should be presumed that Szczepiński made the assumption that the term σ_r/ρ is much smaller than σ_z/ρ and that he omitted it in his equilibrium equation (2.116). On the free surface of the specimen, this assumption is true for any ρ. However, inside the sample, the correctness of this assumption should be verified utilising data obtained from the numerical simulation. Its detailed description is placed in Chap. 3 for three different materials applied in the tensile test. The ratios σ_z/ρ and σ_r/ρ were determined for three advanced stages of deformation of different materials. It turned out that the ratio σ_z/ρ to σ_r/ρ is more than three in the surrounding of the sample axis. Therefore, to a certain approximation, the assumption made by Szczepiński can be accepted.

Derivation of the Szczepiński Formula

Let us return to the formula derivation for the average normalised axial stress according to Szczepiński. During the transformation of Eq. 2.116, he took the same simplifying assumptions (2.42) and (2.49) which Bridgman made. Taking into account these conditions and proceeding analogous to the Bridgman approach (see Eqs. 2.50–2.51), i.e. assuming that the yield stress is constant in the plane of the minimal specimen section and utilising the relation (2.54), we have:

$$\frac{\sigma_z}{\rho} + \frac{d\sigma_z}{dr} = 0 \quad \text{for } z = 0. \tag{2.128}$$

Simultaneously, Szczepiński assumed following Davidenkov and Spiridonova that the curvature radii of the longitudinal stress trajectories are connected with the curvature radius of the contour by relation (2.101). Utilising in Eq. 2.128 the relationship (2.101), we have:

$$\frac{r\sigma_z}{Ra} + \frac{d\sigma_z}{dr} = 0 \quad \text{for } z = 0, \tag{2.129}$$

or

$$\frac{d\sigma_z}{\sigma_z} = -\frac{r\,dr}{Ra} \quad \text{for } z = 0, \tag{2.130}$$

with the boundary condition (2.76).

Finally, after integration, we obtain:

$$\frac{\sigma_z}{k} = \exp\left(\frac{a^2 - r^2}{2aR}\right). \tag{2.131}$$

Utilising the formula (2.79) for the average normalised axial stress in the minimum section plane, we have:

$$\frac{\bar{\sigma}_z}{k} = \frac{2}{a^2} \int_0^a \left(\exp\left(\frac{a^2 - r^2}{2aR}\right)\right) r\,dr. \tag{2.132}$$

Finally, after integration, this formula takes the form:

$$\frac{\bar{\sigma}_z}{k} = \frac{2R}{a} \left[\exp\left(\frac{a}{2R}\right) - 1\right]. \tag{2.133}$$

In order to derive the formula for the normalised radial stress in the minimal section plane, we will apply in the equilibrium equation (2.116) the assumption (2.42) and the condition (2.49) as well as the relation (2.101) to obtain:

$$\frac{r(\sigma_r + k)}{Ra} + \frac{d\sigma_r}{dr} = 0, \tag{2.134}$$

with the boundary condition (2.75). Determining the integral in the form

$$\int_{\sigma_r}^{0} \frac{d\sigma_r}{\sigma_r + k} = -\int_r^a \frac{r}{Ra}\,dr, \tag{2.135}$$

we obtain the formula for the normalised radial stress in the plane of the minimal section in the form:

$$\frac{\sigma_r}{k} = \exp\left(\frac{a^2 - r^2}{2aR}\right) - 1. \tag{2.136}$$

Taking into account Eq. 2.42, the formula for the normalised circumferential stress has the same form as for the normalised radial stress according to Eq. 2.136.

Let us additionally consider the equilibrium equation derived by Szczepiński (2.128) and insert it into the formula for the curvature radius of the longitudinal stress trajectories determined by Bridgman (2.73) to obtain:

$$\frac{d\sigma_z}{\sigma_z} = -\frac{2r\,dr}{a^2 + 2Ra - r^2} \quad \text{for } z = 0, \tag{2.137}$$

with the boundary condition (2.76). Finally, after integration, we have:

$$\frac{\sigma_z}{k} = 1 + \frac{a^2 - r^2}{2Ra}. \tag{2.138}$$

It is easy to notice that the obtained formula corresponds to the one derived by Davidenkov and Spiridonova (2.105). Therefore, inserting the relation for the curvature radius of the longitudinal stress trajectory determined by Bridgman (2.73) in the equilibrium equation derived by Szczepiński (2.128) leads to the Siebel–Davidenkov–Spiridonova formula!

Let us consider the equilibrium equation obtained from both approaches in the form of Eq. 2.121 and solve it utilising the simplifying assumptions (2.42) as well as (2.49) and the derived relation (2.54). Such simplifications were applied in Bridgman, Siebel, Davidenkov, Spiridonova and Szczepiński approaches. Therefore, after utilisation of simplifications (2.42), (2.49) and (2.54) in Eq. 2.121, we obtain:

$$\frac{k}{\rho} + \frac{d\sigma_z}{dr} = 0, \tag{2.139}$$

which overlaps with the equilibrium equation (2.70).

To sum up the considerations included in this subsection, it should be noticed that probably Szczepiński realised that the simplified equilibrium equation of the form (2.121) follows the equation obtained by Bridgman, Siebel and Davidenkov–Spiridonova. Additionally, knowing of the inaccuracy of the Bridgman formula, Szczepiński certainly tried to improve this solution by taking into account that $\sigma_r < \sigma_z$! We cannot judge his manner, but as it was presented above, this simplification is justified.

2.7 Formulae Derivation in Malinin and Petrosjan Approach

During derivation of the formula for the average normalised axial stress in the plane of the minimal section, Malinin and Petrosjan [9] considered both equilibrium equations (2.12) and (2.13) in a certain surrounding of the minimal section. After utilisation of the simplifying assumption (2.42) in Eq. 2.12 and multiplying it by r as well as differentiating it with respect to r, we obtain:

$$\frac{\partial}{\partial r}\left(r\frac{\partial\sigma_r}{\partial r}\right) + \frac{\partial^2}{\partial r\partial z}(r\tau_{zr}) = 0. \tag{2.140}$$

However, after differentiating with respect to z and multiplying with r, Eq. 2.13 takes the form:

$$\frac{\partial^2}{\partial r\partial z}(r\tau_{zr}) + r\frac{\partial^2\sigma_z}{\partial z^2} = 0. \tag{2.141}$$

Subtracting both sides of Eq. 2.141 from (2.140), we obtain:

$$\frac{\partial}{\partial r}\left(r\frac{\partial \sigma_r}{\partial r}\right) = r\frac{\partial^2 \sigma_z}{\partial z^2}. \tag{2.142}$$

Assuming that the yield stress is constant in the plane of the minimum specimen section and taking into account the relation (2.54), we have:

$$\frac{\partial}{\partial r}\left(r\frac{\partial \sigma_z}{\partial r}\right) = r\frac{\partial^2 \sigma_z}{\partial z^2}. \tag{2.143}$$

Subsequently, Malinin and Petrosjan assumed that the value of the axial stress can be presented in the form of a product of two functions:

$$\sigma_z = RZ, \tag{2.144}$$

whereby the first one is a function of r, i.e. $R = R(r)$, and the second one is a function of z, i.e. $Z = Z(z)$, as well as both of them are positive. Let us insert the relation (2.144) into (2.143) to obtain:

$$Z\frac{\partial R}{\partial r} + Zr\frac{\partial^2 R}{\partial r^2} = rR\frac{\partial^2 Z}{\partial z^2}. \tag{2.145}$$

Taking into account that the function Z has its maximum at $z = 0$ so that the value Z'' is negative in the surrounding of $z = 0$, we conclude that:

$$\frac{Z''}{Z} = -\lambda^2 + O(z^2) \quad \text{for } z \to 0. \tag{2.146}$$

Malinin and Petrosjan assumed that the relation (2.146) is the identity. It indeed means that the function in the neck surrounding behaves as the following trigonometric relation:

$$Z(z) = Z_0 \cos(\lambda z). \tag{2.147}$$

Let us notice that these two positive coefficients Z_0 and λ are not known at the moment and must be later estimated.

Under consideration of the assumption (2.146), Eq. 2.145 transforms to the form:

$$\frac{dR}{dr} + r\frac{d^2 R}{dr^2} = -\lambda^2 rR. \tag{2.148}$$

Making a substitution of the variable for the function R from r to $\xi = \lambda r$, i.e. $R(r) \to R(\lambda r)$, we get the differential equation for the function $R(\xi)$ in the form:

$$\xi^2 \frac{d^2 R}{d\xi^2} + \xi \frac{dR}{d\xi} + \xi^2 R = 0. \tag{2.149}$$

The differential equation (2.149) represents the Bessel differential equation of zero order. The general integral of this equation is the linear combination of Bessel and Neumann functions of zero order. Taking into account that the Neumann function of the zero order tends to infinity at $r = 0$, we omit it in the solution. Therefore, the axial stress takes the form:

$$\sigma_z(r, z) = Z_0 \cos(\lambda z) J_0(\lambda r),\qquad(2.150)$$

and in the plane of the minimum section it is simplified to:

$$\sigma_z(r, 0) = Z_0 J_0(\xi).\qquad(2.151)$$

Taking into account the relation for the average axial stress $\bar{\sigma}_z$ in a plane parallel to the minimum section according to Eq. 2.79, which means in the nearest surrounding of the minimum section, the tensile force F according to Eq. 2.82 is obtained as:

$$F = 2\pi \int_0^{a_z} \sigma_z r \, dr,\qquad(2.152)$$

where $a_z = a_z(z)$ is the radius of the neck cross section at a distance z from the plane of the minimal section. Obviously at $z = 0$, we have $a_z = a$ (Fig. 2.1).

Let us twice differentiate the formula (2.152) with respect to z, bearing in mind that F takes the same value in each section, we obtain after the first differentiation:

$$0 = \frac{\partial a_z}{\partial z} \sigma_z(a_z, z) a_z + \int_0^{a_z} \frac{\partial \sigma_z(r, z)}{\partial z} r \, dr,\qquad(2.153)$$

and after the second one:

$$0 = \frac{\partial^2 a_z}{\partial z^2} \sigma_z(a_z, z) a_z + \frac{\partial a_z}{\partial z} \frac{\partial \sigma_z}{\partial r} (a_z, z) \frac{\partial a_z}{\partial z} a_z + \frac{\partial a_z}{\partial z} \sigma_z(a_z, z) \frac{\partial a_z}{\partial z}$$

$$+ \frac{\partial a_z}{\partial z} \frac{\partial \sigma_z}{\partial z} (a_z, z) a_z + \int_0^{a_z} \frac{\partial^2 \sigma_z}{\partial z^2} (r, z) r \, dr.\qquad(2.154)$$

Taking into account the symmetry of the tension sample, the function $a_z(z)$ (Fig. 2.1) takes the form:

$$a_z(z) = a + d_1 z^2 + d_2 z^4 + O(z^6),\qquad(2.155)$$

where the sample radius in the plane of the minimal section a and coefficients d_1, d_2 change during deformation. Therefore, the first derivative $a_z'(z) = 2d_1 z + 4d_2 z^3 + O(z^5)$ as $z \to 0$ takes a value of zero and the second one is $a_z''(0) = 2d_1$. Utilising the formula for the curvature radius according to Eq. 2.65 in the form $\rho = [1 + a_z'(z)^2]^{3/2}/a_z''(z)$, we obtain at $z = 0$:

$$\rho|_{z=0} = R = \frac{1}{2d_1}. \qquad (2.156)$$

Let us notice that we have at $z = 0$

$$\frac{\partial \sigma_z}{\partial z} = 0, \quad \frac{\partial a_z}{\partial z} = 0, \quad \frac{\partial^2 a_z}{\partial z^2} = \frac{1}{R}. \qquad (2.157)$$

Taking into account relations (2.157) and (2.54) in Eq. 2.154, we obtain:

$$\int_0^a \frac{\partial^2 \sigma_z}{\partial z^2} r \, dr = -\frac{ka}{R}, \qquad (2.158)$$

where in addition the boundary condition in the form of Eq. 2.76 was used. Integrating Eq. 2.142 and comparing it to formula (2.158) as well as utilising the relation (2.54), we have:

$$\left. \frac{\partial \sigma_z}{\partial r} \right|_{r=a} = -\frac{k}{R}. \qquad (2.159)$$

Comparing the relation (2.151) and the boundary condition (2.76), we obtain:

$$Z_0 J_0(\xi_1) = k, \qquad (2.160)$$

where $\xi_1 = \lambda a$. Calculating the derivative with respect to r from relationship (2.151) with the use of Eq. 2.159, we have:

$$Z_0 \lambda J_0'(\xi_1) = -\frac{k}{R}. \qquad (2.161)$$

Inserting Eq. 2.151 into (2.152), we obtain:

$$\lambda^2 F = 2\pi Z_0 \int_0^{\xi_1} J_0(\xi)\xi \, d\xi. \qquad (2.162)$$

Determining the coefficient Z_0 from Eqs. 2.160 and 2.161, comparing the obtained formulae and utilising in addition the relation linking the Bessel functions in the form $J_0'(\xi) = -J_1(\xi)$ as well the replacement $\xi_1 = \lambda a$, we finally obtain after all transformations

$$\frac{a}{R} = \xi_1 \frac{J_1(\xi_1)}{J_0(\xi_1)} \equiv \Theta(\xi_1). \qquad (2.163)$$

This equation allows to find the constant λ by known values. To this aim, the inverse function of $\Theta(\xi_1)$ should be determined. Let us notice that the function $\Theta(\xi_1)$ defined on the right hand side of this equation is monotonic. Its diagram is presented in the Fig. 2.6.

Fig. 2.6 Diagram of function $\Theta(\xi_1)$ according to Eq. 2.163

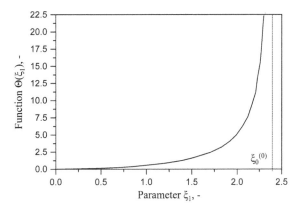

Let us notice that the asymptote of the function is $\xi_1 = \xi_0^{(0)} \approx 2.404825558$ which is a result of the fact that the Bessel function J_0 has the first root in this point. Therefore, the inverse function exists and it can be written in an analytical form. From formula (2.163), ξ_1 can be determined in the form:

$$\xi_1 = \Theta^{-1}\left(\frac{a}{R}\right), \tag{2.164}$$

as well as the constant λ in the form:

$$\lambda = \frac{1}{a}\Theta^{-1}\left(\frac{a}{R}\right). \tag{2.165}$$

Taking into account in Eq. 2.162 the relation (2.160) and in addition the integral from the expression of the Bessel function in the form $\int_0^{\xi_1} J_0(\xi)\xi \, d\xi = \xi_1 J_1(\xi_1)$ and (2.82), as well as the replacement $\xi_1 = \lambda a$, after transformations, we finally obtain the formula for the average normalised axial stress in the minimum section plane as:

$$\frac{\bar{\sigma}_z}{k} = \frac{2J_1(\xi_1)}{\xi_1 J_0(\xi_1)}. \tag{2.166}$$

The final relation can be easily transformed with reference to (2.163) and (2.164) to the form:

$$\frac{\bar{\sigma}_z}{k} = \frac{2}{\left[\Theta^{-1}\left(\frac{a}{R}\right)\right]^2}\frac{a}{R}. \tag{2.167}$$

This means indeed that the Malinin–Petrosjan formula is based on the same values measurable from the experiment as the Bridgman and Siebel–Davidenkov–Spiridonova equation and also depends only on their ratio.

Fig. 2.7 Diagram of the inverse function $\Theta^{-1}(\xi)$, the asymptotical expression (2.169) and the approximate function (2.170)

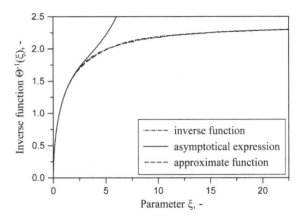

Determining the asymptotical behaviour of $J_1(\xi_1)/J_0(\xi_1)$, the relation (2.163) takes the form:

$$\Theta(\xi_1) = \frac{1}{2}\xi_1^2 + \frac{1}{16}\xi_1^4 + \frac{1}{96}\xi_1^6 + O(\xi_1^8) \quad \text{for } \xi_1 \to 0. \tag{2.168}$$

Determining the behaviour of the inverse function from Eq. 2.168, we have:

$$\Theta^{-1}(\xi) = \sqrt{2\xi}\left(1 - \frac{\xi}{8} + \frac{5\xi^2}{384} + O(\xi^3)\right) \quad \text{for } \xi \to 0. \tag{2.169}$$

The diagram of the inverse function is presented in Fig. 2.7. As it follows from the earlier considerations, $\Theta^{-1}(\xi) \to \xi_0^{(0)}$ for $\xi \to \infty$. Taking into account that the function cannot be written in the form of elementary functions, we decided to approximate it in the whole range of its values ($0 \le \xi < \infty$) in the form:

$$\Theta_*(\xi) = \sqrt{2\xi}\left(1 - \frac{\xi}{8} + \frac{5\xi^2}{384} + \eta_1\xi^3 + \xi_0^{(0)}\eta_2\xi^4\right)\Big/\left(1 + \sqrt{2\xi}\eta_2\xi^4\right), \tag{2.170}$$

where constants $\eta_1 = -0.001684$ and $\eta_2 = 0.00052196$.

It is worth noticing that the function $\Theta_*(\xi)$ behaves asymptotical at $\xi \to 0$ and $\xi \to \infty$ in the same manner as $\Theta^{-1}(\xi)$ and can be used in order to determine the inverse function $\Theta^{-1}(\xi)$. In addition to the diagram of function $\Theta^{-1}(\xi)$, Fig. 2.7 includes the asymptotical expression (2.169) and the approximate form of the inverse function $\Theta_*(\xi)$.

It should be noticed that the asymptotical expression in the form of Eq. 2.169 approximates the inverse function with good accuracy only for values of the parameter ξ up to approximately 2.4. After this value is exceeded, a rapid decrease of the accuracy follows. However, the approximate function in the form of Eq. 2.170 accurately approximates the inverse function and the highest error of approximation is lower than 1%.

Taking into account the mentioned earlier approximate function in the engineering approximation, this formula can be transformed without any accuracy loss by the approximate formula:

$$\frac{\bar{\sigma}_z}{k} = \frac{2}{\Theta_*^2\left(\frac{a}{R}\right)}\frac{a}{R},$$
(2.171)

or

$$\frac{\bar{\sigma}_z}{k} = \frac{\left(1 - \sqrt{2}\eta_2\left(\frac{a}{R}\right)^{9/2}\right)^2}{\left(1 - \frac{a}{8R} + \frac{5}{384}\left(\frac{a}{R}\right)^2 + \eta_1\left(\frac{a}{R}\right)^3 + \xi_0^{(0)}\eta_2\left(\frac{a}{R}\right)^4\right)^2},$$
(2.172)

where the values of constants η_1, η_2 and $\xi_0^{(0)}$ are determined above. It is worth noticing that at small values of the parameter a/R and taking advantage of the asymptotical formula for the inverse function, we have:

$$\frac{\bar{\sigma}_z}{k} = 1 + \frac{a}{4R} + O\left(\frac{a}{R}\right)^2,$$
(2.173)

which overlaps with the asymptotical behaviour following from the Bridgman and Siebel–Davidenkov–Spiridonova formulae.

In order to determine the formula for the normalised axial stress in the minimal section plane, we utilise Eqs. 2.151 and 2.160 to obtain:

$$\frac{\sigma_z}{k}(r,0) = \frac{J_0(\lambda r)}{J_0(\lambda a)},$$
(2.174)

where the value of coefficient λ is determined by the formula (2.165). However, the formula for the remaining two normalised stresses can be obtained from the relations (2.49) and (2.42) as

$$\frac{\sigma_r}{k}(r,0) = \frac{\sigma_\theta}{k}(r,0) = \frac{J_0(\lambda r)}{J_0(\lambda a)} - 1.$$
(2.175)

2.8 New Formula Derivation by Generalisation of the Relation for the Curvature Radius of Longitudinal Stress Trajectory from the Classical Approaches

A new formula for the average normalised axial stress in the minimum cross section plane is derived in this subsection as a result of generalising the classical formulae for the curvature radius of the longitudinal stress trajectories.

Let us notice that the relations (2.73) and (2.101) can be presented in a general form [5] as

$$\rho = \frac{Ra}{rG'(r^2/a^2)}.$$ (2.176)

Let us assume according to the presumption made by Bridgman and Siebel–Davidenkov–Spiridonova that the function $\rho(r)$ is strictly decreasing:

$$\rho'(r) < 0 \quad \text{for } 0 < r < a,$$ (2.177)

in the manner written by the relations (2.102). These relations can be transformed to the function $G'(t)$ to obtain:

$$Ra\{[rG'(r^2/a^2)]^{-1}\}' < 0 \quad \text{for } 0 < r < a,$$ (2.178)

$$\lim_{r \to 0} rG'(r^2/a^2) = 0,$$ (2.179)

and:

$$G'(1) = 1 \text{ and } G'(t) > 0 \quad \text{for } 0 < t \le 1.$$ (2.180)

Let us consider now the condition (2.178), which can be rewritten in the equivalent form:

$$\frac{-[rG'(r^2/a^2)]'}{[rG'(r^2/a^2)]^2} < 0,$$ (2.181)

which results in the relation:

$$[rG'(r^2/a^2)]' > 0.$$ (2.182)

Calculating the integral from the expression (2.182) in the interval $(0, r)$ and taking into account Eq. 2.179, we have the condition:

$$rG'(r^2/a^2) > 0,$$ (2.183)

which indeed overlaps with $(2.180)_2$.

Comparing the relations (2.101) and (2.176), we obtain for the Siebel–Davidenkov–Spiridonova approach:

$$G'_{S-D-S}(t) = 1 \text{ and } G_{S-D-S}(t) = t + c.$$ (2.184)

However, from Eqs. (2.73) and (2.176) for the Bridgman approach, we have:

$$G'_{Br}(t) = \frac{2Ra}{a^2 + 2aR - a^2t} \text{ and } G_{Br}(t) = -2\frac{R}{a}\ln\left(1 + 2\frac{R}{a} - t\right) + c.$$ (2.185)

In these relationships, the constant c is yet optional. In the further part of this section, we will point to the manner of its determination (see Eq. 2.190).

In order to determine the stress σ_z, we utilise the equilibrium equation in the form (2.70) from Sect. 2.3.1, which together with Eq. 2.176 creates the expression:

$$d\sigma_z = -\frac{kG'(r^2/a^2)r\,dr}{aR}. \tag{2.186}$$

The boundary condition on the neck contour (2.76) from Sect. 2.3.1 allows to transform Eq. 2.186 in the form:

$$\int_{\sigma_z}^{k} d\sigma_z = -\frac{k}{aR}\int_{r}^{a} G'(r^2/a^2)r\,dr, \tag{2.187}$$

which, after integration, takes the form:

$$k - \sigma_z = -\frac{ka}{2R}\left[G(1) - G(r^2/a^2)\right]. \tag{2.188}$$

Finally, the formula for the axial stress can be written as

$$\sigma_z = -\frac{ka}{2R}G(r^2/a^2) + \frac{ka}{2R}G(1) + k. \tag{2.189}$$

Separating constants from Eq. 2.189 and comparing them to zero, as well as taking advantage of the fact that the function $G(t)$ is determined with an accuracy to the constant (see Eqs. 2.184 and 2.185), we can assume that:

$$\frac{ka}{2R}G(1) + k = 0. \tag{2.190}$$

Subsequently, we will obtain the following form of function $G(t)$ for the classical solutions:

$$G_{S-D-S}(t) = t - \frac{2R}{a} - 1, \quad G_{Br}(t) = \frac{2R}{a}\left(\ln\frac{2R}{a + 2R - at} - 1\right). \tag{2.191}$$

Let us notice that the functions $G(t)$ and $G'(t)$ are dependent on the variable t and additionally dependent on the dimensionless parameter R/a:

$$G(t) = G(t, R/a). \tag{2.192}$$

Returning to our considerations, Eq. 2.189 can be written in the form:

$$\sigma_z = -\frac{ka}{2R}G(r^2/a^2). \tag{2.193}$$

It is worth mentioning that the condition $(2.180)_2$ points at monotonicity of the stress σ_z as a function of r in the minimum section plane. So the monotonicity condition of function $\rho(r)$ is equivalent to the monotonicity of $\sigma_z(r)$ presuming that k is constant in this section.

The average axial stress $\bar{\sigma}_z$ in the minimum section plane is given by formula (2.79) from Sect. 2.3.1:

$$\bar{\sigma}_z = \frac{2}{a^2} \int_0^a \sigma_z r \, dr = -\frac{k}{aR} \int_0^a G(r^2/a^2) r \, dr. \qquad (2.194)$$

We make the auxiliary replacement

$$F'(t) = G(t), \qquad (2.195)$$

after inserting it into Eq. 2.190, we obtain:

$$\frac{ka}{2R} F'(1) + k = 0. \qquad (2.196)$$

Let us determine now $F'(1)$ in the form:

$$F'(1) = -2\frac{R}{a}. \qquad (2.197)$$

Utilising Eq. 2.195 and the replacement $t = r^2/a^2$, Eq. 2.194 transforms to the form:

$$\bar{\sigma}_z = -\frac{k}{aR} \int_0^a F'(r^2/a^2) r \, dr = -\frac{ka}{2R} [F(1) - F(0)]. \qquad (2.198)$$

Let us notice that the value of function F is determined with an accuracy to the constant, which reduces itself in the square bracket. Thus, it can be always assumed that

$$F(0) = 0, \qquad (2.199)$$

and then Eq. 2.198 has the form:

$$\bar{\sigma}_z = -\frac{ka}{2R} F(1). \qquad (2.200)$$

It is worth writing here the forms of function $F(t)$ for the classical solutions as

$$F_{S-D-S}(t) = \frac{t^2}{2} - \frac{2R}{a} t - t,$$

$$F_{Br}(t) = \frac{2R}{a} \left\{ \left(1 + \frac{2R}{a} - t\right) \left[\ln\left(1 + \frac{2R}{a} - t\right) - 1\right] - t + t\ln\left(\frac{2R}{a}\right) \right\}. \qquad (2.201)$$

Let us notice that similarly to functions $G(t)$ and $G'(t)$, also the function $F(t)$ depends on the dimensionless parameter R/a:

$$F(t) = F(t, R/a). \qquad (2.202)$$

Therefore, the expression (2.200) depends only on the ratio R/a.

We will show that quite simple functions $G(t)$ can be chosen to satisfy the condition (2.180) and finally differing from the results obtained on the base of the Bridgman as well as the Siebel–Davidenkov–Spiridonova formula.

Let us consider a function of the form:

$$G'(t) = \beta + t^{-\frac{\alpha}{2}}(1 - \beta) \quad \text{for} \ -\infty < \alpha < 1. \tag{2.203}$$

It is obvious that $G'(1) = 1$. Let us notice that for the following combination of parameters as $\beta = 1$ and $\alpha \in (-\infty, 1)$ or $\alpha = 0$ as well as any $\beta \in R$, we get the relation given by Siebel–Davidenkov–Spiridonova. Since the second condition of (2.180) was satisfied, it is necessary to assume that:

$$\begin{aligned} \beta \leq 1, \quad & 0 < \alpha < 1, \\ 0 < \beta, \quad & -\infty < \alpha < 0. \end{aligned} \tag{2.204}$$

Let us return to Eq. 2.203, from which after integration we obtain:

$$G(t) = \beta t + \frac{2(1 - \beta)}{2 - \alpha} t^{\frac{2-\alpha}{2}} + c = F'(t). \tag{2.205}$$

Integration of the Eq. 2.205 gives:

$$F(t) = \frac{\beta t^2}{2} + \frac{4(1 - \beta)}{(2 - \alpha)(4 - \alpha)} t^{\frac{4-\alpha}{2}} + ct + c_1. \tag{2.206}$$

Taking advantage from the relation (2.197), constant c can be calculated from the relation (2.205) in the form:

$$c = -2\frac{R}{a} - \beta - \frac{2(1 - \beta)}{2 - \alpha}. \tag{2.207}$$

Inserting Eq. 2.207 into 2.206 with application of the relation (2.200), we finally obtain the formula for the average normalised axial stress in the minimal section plane in the form:

$$\frac{\bar{\sigma}_z}{k} = 1 + \frac{a}{4R} + \frac{a(1 - \beta)\alpha}{4R(4 - \alpha)}, \tag{2.208}$$

were parameters α and β are to be determined from the condition (2.204).

It should be noted here that the limitations for the parameters α and β in Eq. 2.204 were estimated based on the monotonicity assumption of the curvature radius distribution of the longitudinal stress trajectory in the minimal section plane. This assumption was used by Siebel, Bridgman and Daviedenkov–Spiridonova, but it may turn out to be not true at the end. Therefore, one should not give significance to these limitations. It is worth noticing that the formula (2.208) has the form of the Siebel expression (2.115) and allows to find the unknown value of the parameter n as a function of these two new parameters α, β in the following

$n = [4(4 - \alpha)]/[4 - \alpha + \alpha(1 - \beta)] - 3$. As it will be presented in Sect. 3.4, the assumed values $\alpha = \beta = 0.5$ will lead to a more accurate formula in comparison with the discussed classical expressions. For such values of the parameters α and β, we obtain $n = 11/15$ and the assumed value of $n = 1$ by Siebel may not be the most appropriate solution.

2.9 New Formula Derivation on the Base of Other Set of Assumptions

The deformation theory of plasticity will be consistently applied in the following considerations because a homogeneous and monotonic deformation occurs up to the moment of neck creation. Simultaneously, one can expect that the error connected with the application of the deformation theory will be lower in comparison with results obtained on the base of the more general plastic flow theory as long as the neck is not deep. On the other hand, such a consideration considerably simplifies the analysis. Additionally, the Huber–Mises yield condition will be applied in the whole analysis.

The displacements along r and z axes measured in Euler coordinates are assumed in the form:

$$
\begin{aligned}
u_r &= B_0(r) + B_2(r)z^2 + B_4(r)z^4 + O(z^6), \\
u_z &= A_1(r)z + A_3(r)z^3 + A_5(r)z^5 + O(z^7).
\end{aligned}
\tag{2.209}
$$

The choice of the functions u_r and u_z is limited by the fact that a small surrounding of the minimum section plane is considered. Obviously, displacements u_r, u_z should be smooth enough and represented by even and odd function, respectively. Simultaneously, the functions $A_i(r)$ and $B_j(r)$ are smooth with respect to the variable r.

Taking advantage of the equations of deformation theory of plasticity (see Sect. 2.1.4), one can formulate:

$$
\varepsilon_r = \Psi(\sigma_r - \sigma^0), \quad \varepsilon_z = \Psi(\sigma_z - \sigma^0), \quad \varepsilon_\theta = \Psi(\sigma_\theta - \sigma^0), \quad \varepsilon_{rz} = \Psi\tau_{rz}, \tag{2.210}
$$

where $\sigma^0 = (\sigma_r + \sigma_z + \sigma_\theta)/3$ is the hydrostatic pressure and Ψ is an unknown function, which has to be just determined and which is connected with the yield stress k (for example for the Huber–Mises yield condition by the relation (2.29)). Additionally, the condition of material incompressibility is used in the form:

$$
\varepsilon_r + \varepsilon_z + \varepsilon_\theta = 0. \tag{2.211}
$$

The natural boundary conditions of the problem can be stated in the following form:

$$\tau_{rz}|_{r=0} = 0, \quad \tau_{rz}|_{z=0} = 0, \quad \sigma^{(n)}|_{r=a_z(z)} \equiv 0, \quad u_z(r,0) = 0, \quad u_r(0,z) = 0.$$

$$(2.212)$$

During the experiment we determine the tensile force F which is measured at the ends of the sample. Additionally, as a result of the geometry measurement of the deformed specimen, we gain its radius a in the minimum section plane. Based on the approximation of the neck contour shape, we obtain then the function $a_z(z)$ (Fig. 2.1). Utilising the form of function $a_z(z)$ the curvature radius R in the minimum section plane (see Sect. 2.7) can be calculated. The obtained data set leads to the following relations:

$$u_r(a,0) = a - a_0, \quad \rho(a,0) = R, \quad \bar{\sigma}_z|_{z=0} = \frac{F}{\pi a^2}. \tag{2.213}$$

The relationship which links the position of the points on the free surface with their initial coordinates (Fig. 2.1) has the form:

$$r = a_0 + u_r(r,z) \quad \text{at } r = a_z(z). \tag{2.214}$$

The last Eqs. 2.213 and 2.214 will be utilised as the additional boundary conditions and will enable to determine the functions $A_i(r)$ and $B_j(r)$. However, the condition $(2.212)_4$ is always satisfied for the sake of Eq. 2.209 and the condition $(2.212)_5$ requires to assume the value of function $B_j(r)$ in the form $B_0(0) = B_2(0) = B_4(0)$. On the other hand, the conditions $(2.212)_1$, $(2.212)_2$ can be replaced, after utilisation of relation $(2.210)_4$, by

$$\varepsilon_{rz}|_{r=0} = 0, \quad \varepsilon_{rz}|_{z=0} = 0. \tag{2.215}$$

Estimating the strain tensor components from the relation (2.24), we have:

$$\varepsilon_z = A_1 + 3A_3 z^2 + 5A_5 z^4 + O(z^6), \tag{2.216}$$

$$\varepsilon_r = B'_0 + B'_2 z^2 + B'_4 z^4 + O(z^6), \tag{2.217}$$

$$\varepsilon_\theta = \frac{B_0}{r} + \frac{B_2}{r} z^2 + \frac{B_4}{r} z^4 + O(z^6), \tag{2.218}$$

$$\varepsilon_{rz} = \frac{1}{2}[A'_1 z + A'_3 z^3 + 2B_2 z + 4B_4 z^3] + O(z^5). \tag{2.219}$$

Taking advantage of the material incompressibility condition (2.211), the sum of the functions at the appropriate powers of variable z is equal to zero:

$$B'_0 + \frac{B_0}{r} = -A_1, \quad B'_2 + \frac{B_2}{r} = -3A_3, \quad B'_4 + \frac{B_4}{r} = -5A_5. \tag{2.220}$$

Equations 2.220 are the inhomogeneous ordinary differential equations and their solutions have the form:

$$B_0(r) = -\frac{1}{r}\int_0^r \xi A_1(\xi)\,d\xi, \quad B_2(r) = -\frac{3}{r}\int_0^r \xi A_3(\xi)\,d\xi,$$

$$B_4(r) = -\frac{5}{r}\int_0^r \xi A_5(\xi)\,d\xi. \tag{2.221}$$

Taking into account that $A_i(\xi)$ is a smooth function, after twice integration by parts of the relation $(2.221)_1$, we obtain:

$$B_0(r) = -\frac{r}{2}A_1 + \frac{r^2}{6}A_1' - \frac{1}{r}\int_0^r \frac{\xi^3}{6}A_1''(\xi)\,d\xi. \tag{2.222}$$

The forms of functions $B_2(r)$ and $B_4(r)$ are similar which allows to conclude of their behaviour at $r \to 0$. In addition, there are conditions (2.215) to be satisfied. However, looking at the form of Eq. 2.219, the second one is satisfied automatically and in the first one the sum of functions at the appropriate powers of the variable z is equal to zero, therefore:

$$A_1'(0) + 2B_2(0) = 0, \quad A_3'(0) + 4B_4(0) = 0, \quad A_5'(0) = 0. \tag{2.223}$$

Taking into consideration the earlier presented assumption in the form $B_0(0) = B_2(0) = B_4(0)$, one finally obtains from Eq. 2.223:

$$A_1'(0) = 0, \quad A_3'(0) = 0, \quad A_5'(0) = 0. \tag{2.224}$$

Let us make the assumption that the function which describes the trajectories of the longitudinal stress inside the neck will be denoted as $r = \varphi(z, r)$. The differential equation of the longitudinal stress trajectory in a small surrounding of the minimal section plane ($z \ll 1$) has been derived utilising the geometrical interpretation of the derivative $\tan\psi = r'(z)$ and formulae for stresses in planes perpendicular to one of the principle planes, where the normal creates the angle ψ with the principle axis (2.56) and (2.58) as well as (2.100) in the form:

$$\tan 2\psi = \frac{2\tan\psi}{1 - \tan^2\psi} = \frac{2\tau_{rz}}{\sigma_z - \sigma_r}. \tag{2.225}$$

Finally, after application of the formulae for the deformation theory of plasticity (2.210), the differential equation of the longitudinal stress trajectory takes the form:

$$\frac{\varepsilon_{rz}}{\varepsilon_z - \varepsilon_r} = \frac{\varphi'}{1 - (\varphi')^2} = \varphi' + (\varphi')^3 + O(\varphi')^5, \quad \varphi'(z) \to 0, \quad z \to 0. \tag{2.226}$$

The derivative of the longitudinal stress trajectory is sought in the form:

$$\varphi'(z) = \varphi'(z, r) = zc_1(r) + c_2(r)z^3 + O(z^5) \quad \text{for } z \to 0, \tag{2.227}$$

where $c_1(r)$, $c_2(r)$ are determined from the estimation:

$$\frac{\varepsilon_{rz}}{\varepsilon_z - \varepsilon_r} = z[c_1(r) + c_2(r)z^2] + O(z^5) \quad \text{for } z \to 0, \tag{2.228}$$

taking into consideration (2.216), (2.217) and (2.219). As a result, the functions $c_1(r)$, $c_2(r)$ depend on the same unknown functions $A_i(r)$ and $B_j(r)$. The curvature radius of the longitudinal stress trajectory will be calculated from the formula:

$$\rho = \left.\frac{(1 + r'(z)^2)^{\frac{3}{2}}}{|r''(z)|}\right|_{z=0} = \frac{1}{|c_1(r)|} = 2\left|\frac{A_1 - B_0'}{A_1' + 2B_2}\right|. \tag{2.229}$$

Utilising the condition (2.212)$_3$ (see Fig. 2.1) and the additional assumption that the neck is not deep, we finally obtain for $r = a$:

$$\left.\frac{\varepsilon_{rz}}{\varepsilon_z - \varepsilon_r}\right|_z = \frac{a_z'}{1 - (a_z')^2} \sim a_z'\left(1 + (a_z')^2\right) \quad \text{for } a_z' \to 0 \text{ and } z \to 0, \tag{2.230}$$

where $r = a_z(z)$ is the earlier mentioned function describing the neck contour (Fig. 2.1). Using the information of the form of function $r = a_z(z)$ from Sect. 2.7 and replacing both sides of relation (2.230) as well as comparing coefficients at appropriate powers of the variable z^1 and z^3, we obtain:

$$\left.\frac{A_1' + 2B_2}{A_1 - B_0'}\right|_{r=a} = 4d_1, \tag{2.231}$$

$$\frac{1}{A_1 - B_0'}\left[A_3' + 4B_4 - \frac{3A_3 - B_2'}{A_1 - B_0'}(A_1' + 2B_2)\right]\Bigg|_{r=a} = 16d_1^3 + 8d_2. \tag{2.232}$$

It is worth noticing that both groups of equations, i.e. (2.226)–(2.228) and (2.230)–(2.232), represent the same type of conditions with the known value of the initial specimen radius a_0 and the current radius a, as well as the curvature radius of the deformed sample contour R and the unknown functions $A_i(r)$ and $B_j(r)$. The first group holds inside the neck, while the second one – on the free surface.

Utilising the condition (2.214) and replacing the right-hand side of the relation (2.209)$_1$ and knowing the function describing the neck contour, we have:

$$B_0(a) = a - a_0, \quad \frac{a}{R} = 2a\frac{B_2(a)}{1 - B_0'(a)}, \tag{2.233}$$

for the appropriate power of the variable z^0 and z^2. Let us draw the attention to the fact that Eq. 2.231 responds the natural condition (2.213)$_2$, however, Eq. 2.233$_1$ reflects the condition (2.213)$_1$.

Let us introduce now two new parameters in the form:

$$\delta = \frac{a}{R}, \quad \Lambda = 1 - \frac{a_0}{a}, \quad \text{where } 0 < \delta < \infty \text{ and } -\infty < \Lambda < 0. \tag{2.234}$$

The ratio δ is the independent variable of classical formulae (see formulae (2.81), (2.107) and (2.133)), and it also unequivocally characterises the stage of strain location in the neck surrounding. However, the introduced new parameter Λ characterises the stage of plastic strains gained in the whole specimen. Inserting parameters (2.234) into the relations (2.231), (2.232) and (2.233) we have, respectively:

$$\delta = \frac{a A_1' + 2B_2}{2 A_1 - B_0'}\bigg|_{r=a}, \qquad (2.235)$$

$$\frac{1}{2(A_1 - B_0')}\left[A_3' + 4B_4 - \frac{3A_3 - B_2'}{A_1 - B_0'}\left(A_1' + 2B_2\right)\right]\bigg|_{r=a} = \frac{\delta^3}{a^3} + 4d_2, \qquad (2.236)$$

$$\frac{B_0(a)}{a} = \Lambda, \qquad (2.237)$$

$$\delta = 2a\frac{B_2(a)}{1 - B_0'(a)}. \qquad (2.238)$$

It is worth recalling here from Sect. 2.7 that the coefficients d_1, d_2 occur in the approximating function of the deformed contour (see formula (2.155)), but the coefficient d_1 is linked with the curvature radius of the deformed sample contour R given in relation (2.156). However, the coefficient d_2 is usually not determined because its accuracy estimation (see Eq. 2.155) is a few orders below than that of coefficient d_1 (or R), which in turn also has its limitation regarding the determination accuracy (see Sect. 1.4). Therefore, for estimation of functions B_0 and B_2 at the point $(0, a)$, we have at our disposal the three Eqs. 2.235, 2.237 and 2.238 which are determined in the single point $r = a$. For that sake we are not able to determine them unequivocally and therefore it is necessary to take an equation into account which has not been utilised until now. This issue will be subject of our consideration in the further part of this chapter.

The present considerations will be made for the minimum section plane, i.e. for $z = 0$, where the equilibrium equation (2.12) must be satisfied.

Being based on the equations of the deformation theory of plasticity (2.210) and strains in the form of Eqs. 2.216–2.219, the equilibrium equation (2.12) can be rewritten in the equivalent form:

$$\frac{\partial \sigma_r}{\partial r} + \frac{r^2 A_1'(r) + 2r^2 B_2(r) + 2r B_0'(r) - 2B_0(r)}{2r^2 (A_1(r) - B_0'(r))}(\sigma_z - \sigma_r) = 0, \qquad (2.239)$$

with the boundary condition of the ordinary differential equation:

$$\sigma_r|_{r=a} = 0. \qquad (2.240)$$

Inserting the additional parameter:

$$\gamma(r) = \frac{\varepsilon_r - \varepsilon_\theta}{\varepsilon_z - \varepsilon_r}\bigg|_{z=0} = \frac{B'_0 - \frac{B_0}{r}}{A_1 - B'_0}, \tag{2.241}$$

which occurs in Eq. 2.239, it can be concluded from Eq. 2.210 and the Huber–Mises yield condition that

$$\sigma_z - \sigma_r = \frac{k}{\sqrt{1 + \gamma(r) + \gamma^2(r)}}. \tag{2.242}$$

Finally, the equilibrium equation (2.239) after using the relation (2.242), can be written in the form:

$$\frac{\partial \sigma_r}{\partial r} + f(r)k(r) = 0, \tag{2.243}$$

where

$$f(r) = \frac{r^2 A'_1(r) + 2r^2 B_2(r) + 2r B'_0(r) - 2B_0(r)}{2r^2(A_1(r) - B'_0(r))} \frac{1}{\sqrt{1 + \gamma(r) + \gamma^2(r)}}. \tag{2.244}$$

Integrating Eq. 2.243 with employing the relation (2.244) and the boundary condition according to Eq. 2.240, we obtain:

$$\sigma_r = -\int_a^r \frac{t^2 A'_1(t) + 2t^2 B_2(t) + 2t B'_0(t) - 2B_0(t)}{2t^2(A_1(t) - B'_0(t))} \frac{k}{\sqrt{1 + \gamma(r) + \gamma^2(r)}} \, dt. \tag{2.245}$$

Utilising one more time Eq. 2.242, we finally obtain the distribution of the axial stress in the neck as:

$$\sigma_z - \frac{k}{\sqrt{1 + \gamma(r) + \gamma^2(r)}}$$
$$= -\int_a^r \frac{t^2 A'_1(t) + 2t^2 B_2(t) + 2t B'_0(t) - 2B_0(t)}{2t^2(A_1(t) - B'_0(t))} \frac{k}{\sqrt{1 + \gamma(r) + \gamma^2(r)}} \, dt. \tag{2.246}$$

Let us notice that in the case of adopting the additional assumption utilised by Bridgman, Siebel–Davidenkov–Spiridonova, Szczepiński and Malinin–Petrosjan on the equality of the radial and circumferential stresses together with the associated flow rule, the value of parameter $\gamma(r)$ is equal to zero. Additionally, in all classical approaches the yield stress in the minimum section plane takes a constant value. Taking advantage of these two simplifying assumptions in Eq. 2.246, one can obtain:

$$\frac{\sigma_z}{k} = 1 - \int_a^r \frac{t^2 A_1'(t) + 2t^2 B_2(t) + 2t B_0'(t) - 2B_0(t)}{2t^2 (A_1(t) - B_0'(t))} \, dt. \tag{2.247}$$

Let us determine now the average axial stress in the minimum section plane from this equation in the form:

$$\frac{\bar{\sigma}_z}{k} = \frac{2}{a^2} \int_0^a r \, dr - \frac{2}{a^2} \int_0^a r \int_a^r \frac{t^2 A_1'(t) + 2t^2 B_2(t) + 2t B_0'(t) - 2B_0(t)}{2t^2 (A_1(t) - B_0'(t))} \, dt \, dr, \tag{2.248}$$

which takes after integration the form:

$$\frac{\bar{\sigma}_z}{k} = 1 + \int_0^a \frac{r^2 A_1'(r) + 2r^2 B_2(r) + 2r B_0'(t) - 2B_0(r)}{2a^2 (A_1(r) - B_0'(r))} \, dr. \tag{2.249}$$

In order to solve it, the knowledge about functions B_0 and B_2 is essential. Taking into account the additional assumption that $\gamma = 0$ (see relation (2.241)) the function B_0 is estimated from the equation $B'_0 = B_0/r$, which in turn is satisfied only for a linear function of the form $B_0 = b_{10} r$. In this case, utilising Eq. 2.220$_1$ we obtain the form of function $A_1 = -2b_{10}$ and its derivative $A'_1 = 0$. Therefore, the assumption $\gamma = 0$ together with the remaining conditions (2.235), (2.237) and (2.238) as well as the optional function B_2 lead to the following relations:

$$\delta = a \frac{B_2(a)}{-b_{10}}, \quad b_{10} = \Lambda, \quad \delta = 2a \frac{B_2(a)}{1 - b_{10}}, \tag{2.250}$$

which allow to determine the two contradictory values of function $B_2(a)$ in the form:

$$B_2(a) = -\delta \frac{\Lambda}{a}, \quad B_2(a) = \delta \frac{1 - \Lambda}{2a}. \tag{2.251}$$

It follows that the assumption $\gamma = 0$ leads to incorrect results based on the assumptions made by us and cannot be taken into account in the further considerations. Let us return to the general form of the equation for the distribution of the axial stress in the neck (2.246) bearing in mind that $\gamma \neq 0$ and let us determine from it the average axial stress in the minimum section plane in the form:

$$\bar{\sigma}_z = \frac{2}{a^2} \int_0^a \sigma_z r \, dr = \frac{2}{a^2} \int_0^a \frac{k r \, dr}{\sqrt{1 + \gamma(r) + \gamma^2(r)}}$$

$$- \frac{2}{a^2} \int_0^a r \, dr \int_a^r \frac{t^2 A_1'(t) + 2t^2 B_2(t) + 2B_0'(t) - 2B_0(t)}{2t^2 (A_1(t) - B_0'(t))} \frac{k}{\sqrt{1 + \gamma(r) + \gamma^2(r)}} \, dt.$$

$$\tag{2.252}$$

Integrating the second integral by parts, Eq. 2.252 becomes:

$$
\bar{\sigma}_z = \frac{2}{a^2} \int_0^a \frac{k r \, dr}{\sqrt{1 + \gamma(r) + \gamma^2(r)}}
$$

$$
- \frac{2}{a^2} \frac{r^2}{2} \int_a^r \frac{t^2 A_1'(t) + 2t^2 B_2(t) + 2t B_0'(t) - 2B_0(t)}{2t^2 (A_1(t) - B_0'(t))} \frac{k}{\sqrt{1 + \gamma(t) + \gamma^2(t)}} \, dt \Bigg|_0^a
$$

$$
+ \frac{2}{a^2} \int_0^a \frac{r^2}{2} \frac{r^2 A_1'(r) + 2t^2 B_2(t) + 2r B_0'(r) - 2B_0(r)}{2r^2 (A_1(r) - B_0'(r))} \frac{k}{\sqrt{1 + \gamma(r) + \gamma^2(r)}} \, dr.
$$

$$(2.253)$$

Taking into account that the second term is equal to zero and collecting the common term in front of the bracket, we finally obtain:

$$
\bar{\sigma}_z = \frac{2}{a^2} \int_0^a \frac{k}{\sqrt{1 + \gamma(r) + \gamma^2(r)}} \left[r + \frac{r^2 A_1'(r) + 2r^2 B_2(r) + 2r B_0'(r) - 2B_0(r)}{4(A_1(r) - B_0'(r))} \right] dr.
$$

$$(2.254)$$

Although the assumption of the equality of the radial and circumferential stresses ($\gamma = 0$) is not true in general (see Sect. 3.2), the difference between ε_r and ε_θ is small and considerably less than the difference in the denominator of the parameter γ (2.241). This allows us to simplify partially Eq. 2.254 by changing it by the asymptotic relation with accuracy of $O(\gamma^2)$. Simultaneously, we make an additional assumption, taking into account that the deviation of the function $k(r)$ from the constant value is small and of the same order as γ^2, so that:

$$
k(r) = \bar{k} + \Delta k(r), \quad \text{where } \bar{k} = \frac{2}{a^2} \int_0^a k(r) r \, dr \text{ and } \Delta k(r) = O(\gamma^2) \text{ for } \gamma \to 0.
$$

$$(2.255)$$

The legitimation of this assumption will be discussed in Sect. 3.2.2.2. Taking advantage of both assumptions, Eq. 2.254 transforms to the form:

$$
\frac{\bar{\sigma}_z}{\bar{k}} = \frac{2}{a^2} \int_0^a \left[1 - \frac{1}{2}\gamma + O(\gamma^2) \right] \left[r + \frac{r^2 A_1'(r) + 2r^2 B_2(r) + 2r B_0'(r) - 2B_0(r)}{4(A_1(r) - B_0'(r))} \right] dr.
$$

$$(2.256)$$

Replacing the relation for the parameter γ according to Eq. 2.241 we have with an accuracy of order $O(\gamma^2)$:

$$\frac{\bar{\sigma}_z}{k} = \frac{2}{a^2} \int\limits_0^a \left[1 - \frac{rB'_0 - B_0}{2r(A_1 - B'_0)}\right]\left[r + \frac{r^2A'_1 + 2r^2B_2 + 2rB'_0 - 2B_0}{4(A_1 - B'_0)}\right] dr. \quad (2.257)$$

After further simplification, we obtain:

$$\frac{\bar{\sigma}_z}{k} = \frac{2}{a^2} \int\limits_0^a \left[r + \frac{r^2A'_1 + 2r^2B_2}{4(A_1 - B'_0)}\right] dr. \quad (2.258)$$

In order to solve this relation the knowledge about functions B_0 and B_2 is necessary. As it was mentioned above, there is not enough information to find these values. Therefore, we assume the probably simplest form of a third order polynomial which can be written as

$$B_0(r) = b_{10}r + b_{30}\frac{r^3}{a^2}, \quad B_2(r) = b_{12}r + b_{32}\frac{r^3}{a^2}. \quad (2.259)$$

Taking advantage of the relation (2.220), we get:

$$A_1(r) = -2b_{10} - 4b_{30}\frac{r^2}{a^2}, \quad A_3(r) = -\frac{2}{3}b_{12} - \frac{4}{3}b_{32}\frac{r^2}{a^2}. \quad (2.260)$$

Inserting the relations (2.259) and (2.260) into the conditions (2.235), (2.237) and (2.238), we obtain:

$$\delta = a^2\frac{b_{12} + b_{32} - 4\frac{b_{30}}{a^2}}{-3b_{10} - 7b_{30}}, \quad b_{10} + b_{30} = \Lambda, \quad \delta = \frac{2a^2(b_{12} + b_{32})}{1 - b_{10} - 3b_{30}}. \quad (2.261)$$

It is possible to unequivocally determine the coefficients b_{10}, b_{30} and $b_{12} + b_{32}$ from these relationships. However, we do not have any other conditions for the separation of coefficients b_{12} and b_{32}. From the numerical simulation we succeeded to gain the values of the coefficients in the form:

$$b_{10} = \frac{8\Lambda - \delta - 5\delta\Lambda}{8}, \quad b_{12} = \delta\frac{1 - 6\Lambda}{2a^2}, \quad b_{30} = \delta\frac{1 + 5\Lambda}{8}, \quad b_{32} = \delta\frac{5\Lambda}{2a^2}, \quad (2.262)$$

which simultaneously satisfy the conditions (2.261) with an accuracy of order $O(\delta^2)$.

Taking advantage of functions $A_i(r)$ and $B_j(r)$ in the form of Eqs. 2.259 and 2.260, we obtain the formula for the average normalised axial stress in the minimum section plane as:

$$\frac{\bar{\sigma}_z}{k} = 1 - \frac{b_{32}a^2}{28b_{30}} - \frac{b_{12}a^2}{14b_{30}} + \frac{2}{7} + \frac{3b_{10}b_{32}a^2}{98b_{30}^2}$$

$$+ \frac{3b_{10}}{7b_{30}}\left(\frac{b_{12}a^2}{14b_{30}} - \frac{2}{7} - \frac{3b_{10}b_{32}a^2}{98b_{30}^2}\right) \ln\left|1 + \frac{7b_{30}}{3b_{10}}\right|. \quad (2.263)$$

Finally, after replacement of the parameters b_{ij} based on Eq. 2.262, the relation for the average normalised axial stress takes the form [6]:

$$\frac{\bar{\sigma}_z}{k} = 1 - \frac{5\Lambda}{7(1+5\Lambda)} - \frac{2(1-6\Lambda)}{7(1+5\Lambda)} + \frac{2}{7} + \frac{30\Lambda(8\Lambda - \delta - 5\delta\Lambda)}{49\delta(1+5\Lambda)^2}$$
$$+ \frac{3(8\Lambda - \delta - 5\delta\Lambda)}{7\delta(1+5\Lambda)} \left(\frac{2(1-6\Lambda)}{7(1+5\Lambda)} - \frac{2}{7} - \frac{30\Lambda(8\Lambda - \delta - 5\delta\Lambda)}{49\delta(1+5\Lambda)^2} \right) \ln\left|1 + \frac{7\delta(1+5\Lambda)}{3(8\Lambda - \delta - 5\delta\Lambda)}\right|.$$

$$(2.264)$$

It is worth noticing that the formula (2.264) leads to the same result, i.e. $\bar{k} = \bar{\sigma}_z$, if the parameter δ takes its limit for $\Lambda \to 0$ or if the parameter $\Lambda < 0$ takes its limit for $\delta \to 0$. This limit case is compatible with the expectations.

This formula can be written in its simplest form by applying the replacement:

$$\Gamma = \frac{2(1-6\Lambda)}{7(1+5\Lambda)} - \frac{2}{7} - \frac{30\Lambda(8\Lambda - \delta - 5\delta\Lambda)}{49\delta(1+5\Lambda)^2} \quad \text{and} \quad \Phi = \frac{7\delta(1+5\Lambda)}{3(8\Lambda - \delta - 5\delta\Lambda)}, \quad \text{to obtain:}$$

$$\frac{\bar{\sigma}_z}{k} = 1 - \frac{5\Lambda}{7(1+5\Lambda)} - \Gamma + \frac{\Gamma}{\Phi}\ln|1 + \Phi|. \qquad (2.265)$$

In order to determine the formula for the normalised radial stress in the minimum section plane, it should be utilised the fact that $\gamma(r) \ll 1$ in Eq. 2.245 and replaced the relation for the parameter γ to obtain:

$$\frac{\sigma_r}{k} = -\int_a^r \frac{t^2 A_1'(t) + 2t^2 B_2(t) + 2t B_0'(t) - 2B_0(t)}{2t^2(A_1(t) - B_0'(t))} \left[1 - \frac{t B_0'(t) - B_0(t)}{2t(A_1(t) - B_0'(t))}\right] dt.$$

$$(2.266)$$

Taking advantage of the form of functions $A_i(r)$ and $B_j(r)$ given in Eqs. 2.259 and 2.260, we obtain:

$$\frac{\sigma_r}{k} = \int_a^r t \frac{b_{12}a^2 + b_{32}t^2 - 2b_{30}}{3b_{10}a^2 + 7b_{30}t^2} \left(1 + \frac{b_{30}t^2}{3b_{10}a^2 + 7b_{30}t^2}\right) dt. \qquad (2.267)$$

We proceed in an identical way with the formula for the normalised axial stress in the minimum section plane. After transformation we get from Eq. 2.247:

$$\frac{\sigma_z}{k} = 1 - \frac{r B_0'(r) - B_0(r)}{2r(A_1(r) - B_0'(r))}$$
$$- \int_a^r \left[\frac{t^2 A_1'(t) + 2t^2 B_2(t) + 2t B_0'(t) - 2B_0(t)}{2t^2(A_1(t) - B_0'(t))}\left(1 - \frac{t B_0'(t) - B_0(t)}{2t(A_1(t) - B_0'(t))}\right)\right] dt,$$

$$(2.268)$$

and after utilisation of functions $A_i(r)$ and $B_j(r)$ in the form of Eqs. 2.259 and 2.260, we gain:

$$\frac{\sigma_z}{k} = 1 + \frac{b_{30}r^2}{3b_{10}a^2 + 7b_{30}r^2} + \int_a^r t \frac{b_{12}a^2 + b_{32}t^2 - 2b_{30}}{3b_{10}a^2 + 7b_{30}t^2} \left(1 + \frac{b_{30}t^2}{3b_{10}a^2 + 7b_{30}t^2}\right) dt.$$

(2.269)

However, to determine the normalised circumferential stress in the minimal section plane the assumptions of the deformation theory of plasticity (2.27) should be utilised, to obtain:

$$\sigma_r - \sigma_\theta = (\sigma_z - \sigma_r)\frac{\varepsilon_r - \varepsilon_\theta}{\varepsilon_z - \varepsilon_r} = (\sigma_z - \sigma_r)\gamma(r),$$

(2.270)

from which it can be easily estimated that:

$$\frac{\sigma_\theta}{k} = \frac{\sigma_r}{k} - \frac{\sigma_z - \sigma_r}{k}\frac{rB_0'(r) - B_0(r)}{r(A_1(r) - B_0'(r))},$$

(2.271)

where the formulae determining the radial and axial stresses were given above. Taking advantage of the form of functions $A_i(r)$ and $B_j(r)$ as given in Eqs. 2.259 and 2.260, we finally obtain:

$$\frac{\sigma_\theta}{k} = \frac{\sigma_r}{k} + \frac{\sigma_z - \sigma_r}{k}\frac{2b_{30}r^2}{3b_{10}a^2 + 7b_{30}r^2}.$$

(2.272)

Taking into account that the analytical solution of relations (2.267), (2.269) and (2.272) are very wide, their forms will be not presented here. However, in this case, it is easier to obtain the solutions determining the appropriate integrals in any mathematical program.

References

1. P.W. Bridgman, *Studies in large plastic flow and fracture with special emphasis on the effects of hydrostatic pressure* (Harvard University press, Cambridge, 1964)
2. N.N. Davidenkov, N.I. Spiridonova, Mechanical methods of testing. Analysis of the state of stress in the neck of a tension test specimen. Proc Am Soc Test Mater **46**, 1147–1158 (1947)
3. L. Dietrich, J. Miastkowski, W. Szczepiński, *Limiting capacity of the construction elements (in Polish)* (PWN, Warsaw, 1970)
4. Z. Gabryszewski, J. Gronostajski, *Mechanics of process of plastic forming (in Polish)* (PWN, Warsaw, 1991)
5. M. Gromada, G. Mishuris, A. Öchsner, An attempt to improve the evaluation of mechanical material properties from the axisymmetric tensile test. Israel J Chem **47**, 329–335 (2007)
6. M. Gromada, G. Mishuris, A. Öchsner, On the evaluation of mechanical properties from the axisymmetric tensile test. Arch Metall Mater **55**, 295–300 (2010)
7. R. Hill, *The mathematical theory of plasticity* (Clarendon Press, Oxford, 1950)

8. L.M. Kachanov, *Foundations of the theory of plasticity* (Mir Publishers, Moscow, 1974)
9. N.N. Malinin, J. Rżysko, *Mechanics of materials (in Polish)* (PWN, Warsaw, 1981)
10. M.E. Niezgodziński, T. Niezgodziński, *Strength of materials (in Polish)* (PWN, Warsaw, 2002)
11. W. Olszak, P. Perzyna, A. Sawczuk, *Theory of plasticity (in Polish)* (PWN, Warsaw, 1965)
12. W. Wrona, Ś. Romanowski, *Mathematics for technical courses (in Polish)* (PWN, Warsaw, 1967)
13. E. Siebel, S. Schwaigerer, On the mechanics of the tensile test (in German). Arch Eisenhuttenwes **19**, 145–152 (1948)
14. Standard EN 10002 -1 Metallic materials—Tensile testing—Part 1: Method of test at ambient temperature
15. W. Szczepiński, *Introduction to analysis of plastic forming processes (in Polish)* (PWN, Warsaw, 1967)
16. W. Szczepiński, *Mechanics of plastic flow (in Polish)* (PWN, Warsaw, 1978)
17. W. Szczepiński, *Experimental methods of solid mechanics (in Polish)* (PWN, Warsaw, 1984)

Chapter 3
Formulae Verification for the Flow Curve Determination Due to Numerical Simulation

Abstract The axisymmetric specimen applied in the tensile test is numerically characterised within this chapter. Subsequently, the simplifying assumptions of the classical approaches are verified utilising the results obtained from the numerical simulation and the errors were determined. These investigations can clarify the influence of each particular simplification on the value of the average normalised axial stress in the minimum section plane. Additionally, the errors were estimated which follow from the application of the formula for the true (logarithmic) plastic strain instead of the strain intensity in the flow curve. However, the highest significance, from the point of view of application of particular formulae, had the estimation of the errors created by application of the classical and newly derived formulae to determine the flow curve from the moment of neck formation.

Keywords Analytical and numerical analysis · Simplifying assumptions · Formulae verification

3.1 Description of the Numerical Simulation

The numerical simulation of the tensile test based on axisymmetric samples was carried out using the commercial finite element software MSC.Marc Mentat. The specimen dimensions are chosen in accordance with standard EN 10002-1 [4] and equal to: length 60 mm and diameter 10 mm. Its ends are constrained in the radial direction in order to simulate the fitting in the grip of the tensile testing machine. Because of the sample axial symmetry, only one quarter was modelled, where the axial coordinate varies from 0 to 30 mm and the radial coordinate varies from 0 to 5 mm (Fig. 3.1). In order to increase the accuracy of the calculation, the mesh was refined in the surrounding of the minimum section of the specimen,

M. Gromada et al., *Correction Formulae for the Stress Distribution in Round Tensile Specimens at Neck Presence*, SpringerBriefs in Computational Mechanics, DOI: 10.1007/978-3-642-22134-7_3, © Magdalena Gromada 2011

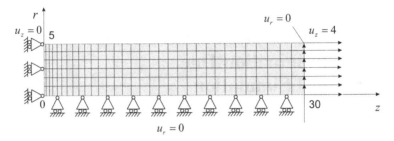

Fig. 3.1 Specimen geometry together with its element division and boundary conditions

where the stress concentration is the highest. Simultaneously, this region is the most interesting from the point of view of the investigated problem.

The model was divided into 64,200 quadrilateral elements with bilinear form function (MSC.Marc element type 10) which is recommended for the analysis of axisymmetric problems [3]. The static boundary conditions take into account the boundary fixation and a load in the form of a displacement which causes the sample tension. In order to ensure the symmetry of the modelled quarter of the sample, it is necessary to assume for all mesh nodes along the axis $z = 0$ a displacements of $u_z = 0$ as well as along the axis $r = 0$ a displacements of $u_r = 0$. Simultaneously, all nodes along $z = 30$ have an assumed displacements of $u_r = 0$. The loading of the sample is realised by a given uniform displacement at the end of specimen, i.e. for $z = 30$, in the form $u_z = 4$ mm for the ideal plastic material (see Fig. 3.1), $u_z = 5$ mm in the case of the linear hardening material and $u_z = 12.5$ mm for the nonlinear hardening material. Obviously, the given values of displacements are gained in the last stage of the incremental calculation. However, in all earlier loading stages the elongations are lower. Figure 3.1 presents a schematic sketch of the axisymmetric sample model applied in the numerical simulation procedure.

It is assumed that the material of all elements of the model is isotropic and reveals elasto-plastic properties. In the regime of elastic strains Young's modulus $E = 210$ GPa and Poisson's ratio $v = 0.3$ are assigned. However, the material behaves in the plastic range according to the Huber-Mises yield condition with an initial yield stress of $k = 200$ MPa.

In the numerical simulation, three model flow curves were applied:

$k(\bar{\varepsilon}^{pl}) = 200$ MPa—ideal plastic material,

$k(\bar{\varepsilon}^{pl}) = 200 + 150\bar{\varepsilon}^{pl}$ MPa—linear hardening material,

$k(\bar{\varepsilon}^{pl}) = 100 + 100 \left(1 + 14.24775\bar{\varepsilon}^{pl}\right)^{0.5}$ MPa—nonlinear hardening material.

The applied calculation algorithm in the program MSC.Marc requires that the strain in the assumed flow curve is given in the form of the true plastic strain $\bar{\varepsilon}^{pl}$. Hence, the form of above presented flow curves was taken as input value.

The material from which the model was prepared for the numerical simulation was chosen in such a manner that the elastic strain regime was small and the

Fig. 3.2 Flow curves applied
in the numerical simulation

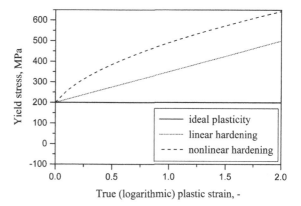

material rapidly reached its yield stress. Only this regime will be useful in the further analytical considerations. However, the flow curves were chosen in such a manner to model quite different behaviour of materials and to have the possibility to analyse the obtained differing results. Based on this idea, the ideal plastic material model with a hardening coefficient equal to zero, as well as a linear hardening material with a constant hardening coefficient of 150 MPa, and a nonlinear hardening material revealing a considerable strain hardening behaviour (see Fig. 3.2) were chosen.

To carry out the numerical simulation with the finite element code MSC.Marc Mentat, appropriate analysis options were chosen which take into account the nonlinearities in regards of both material and geometry. In order to control the solution convergence, the relative force tolerance was taken one order lower (0.001) than the recommended in the User's guide (0.01). Taking into account the specification of the flow curves and the considered stages of deformation of each particular material, the total number of increments in the simulation was different for each material, but repeatable results could be obtained for a given number of increments. For each step of the numerical simulation, all the necessary values were recorded which are required for the further evaluation. Additionally, a Fortran subroutine was written to save the actual coordinates of the specimen in order to evaluate the flow curve based on the deformed shape. The obtained results of the numerical simulation were checked with respect to stability and accuracy.

3.2 Verification of Simplifying Assumptions Applied in Classical Approaches

In this subsection a verification of the basic simplifying assumptions applied during the formulae derivation by Bridgman, Siebel, Davidenkov-Spiridonova, Szczepiński and Malinin-Petrosjan for the average normalised axial stress in the minimum section plane is done by means of numerical simulation as well as

Fig. 3.3 Distribution of
radial and circumferential
stresses in the minimal
section plane of specimen for
three different stages of
deformation of ideal plastic
(**a**) linear hardening (**b**) and
nonlinear hardening
(**c**) materials

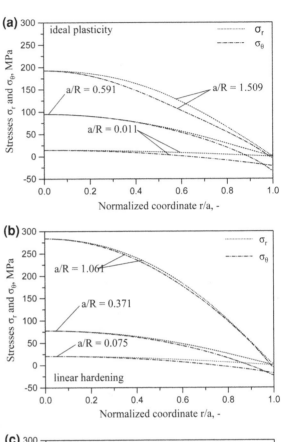

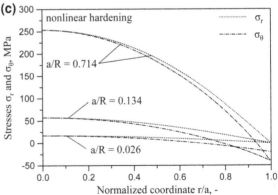

analytical analysis. At the same time, only one of the simplifying assumptions was
analysed and the remaining ones were treated as valid. The considered simplification was investigated by analytical analysis and only necessary data were gained
from the numerical simulation. Such procedure enabled the estimation of the
influence of the considered simplification on the increase of the average

normalised axial stress, and as a result, obtaining the required information which assumption generates the largest error. These considerations were made only for two classical formulae, i.e. Bridgman and Siebel-Davidenkov-Spiridonova and the empirical formula (3.16), which will be presented in Sect. 3.4. In the following investigations, the Szczepiński formula was not under consideration, because, during its derivation, the author utilised the simplified equilibrium equation and the legitimacy of the omission of the term σ_r/ρ by Szczepiński in the applied equilibrium equation was just presented in Sect. 2.6 where this formula was derived. Furthermore, the Malinin and Petrosjan formula was also not taken into account because in the formula for the curvature radius of the longitudinal stress trajectory ρ does not occur as assumption.

3.2.1 Verification of Simplifications by Means of the Numerical Simulation

In this subsection, the verification of the simplifying assumptions utilising data obtained from the numerical simulation is presented. At the beginning, we consider the assumption of the equality of the radial and circumferential stresses. Figure 3.3 presents the distribution of the radial and circumferential stresses in the plane of the minimum section of the specimen. These diagrams were drawn for three chosen stages of deformation and three different hardening behaviours.

From these diagrams it follows that this assumption is not true near the free surface and the stress distribution is only equal in a small surrounding of the sample centre, which is in accordance with the theory.

Let us consider now the assumption of the yield stress constancy in the minimum section plane. To this end, Fig. 3.4 presents the distributions of the Huber-Mises yield stress k together with its average values $\bar{k}$ in the plane of the minimum section of the sample. From Fig. 3.4 it is clearly visible that for small stages of deformation of the tension specimen, the stress distribution slightly differs from the average value. However, the difference is more visible for larger strains. Therefore, it seems that this simplification can turn out to be the most justified, although this observation unnecessary can be totally valid for large strains.

In order to verify the remaining assumption, the values of the curvature radii of the longitudinal stress trajectories were determined from data of the numerical simulation. The determination of the equation describing the distribution $\rho = \rho(r)$ in the surrounding of the minimum section plane of the specimen was presented in Sect. 2.9. In this subsection, we only present the obtained results from its evaluation. Figure 3.5 shows the distribution of the curvature radius of the longitudinal stress trajectory in the minimum section plane of the round sample for the chosen value of deformation of the three different materials. In addition to the results obtained from the numerical simulation, values got from formulae by Bridgman (2.73) and Siebel-Davidenkov-Spiridonova (2.101) are plotted for

Fig. 3.4 Distribution of the
Huber-Mises yield stress
together with its average
value in the minimal section
plane of specimen for three
stages of deformation in the
case of linear hardening
(**a**) and nonlinear hardening
(**b**) materials

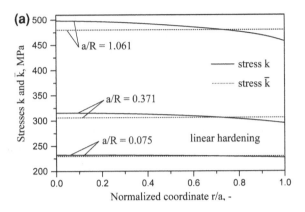

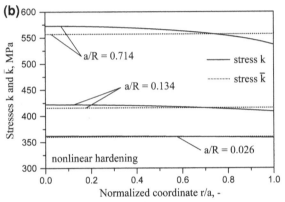

Fig. 3.4 Distribution of the Huber-Mises yield stress together with its average value in the minimal section plane of specimen for three stages of deformation in the case of linear hardening (**a**) and nonlinear hardening (**b**) materials

comparison. Furthermore, the curves obtained from formula (2.176) together with function (2.203) for a parameter set of $\alpha = \beta = 0.5$ (cf. Eq. 3.1) are indicated. The way to identify parameters α and β will be shown in Sect. 3.4, however the results are just shown here.

$$\rho = \frac{2Ra}{r[1 + 1/\sqrt{r/a}]}. \tag{3.1}$$

It is visible in Fig. 3.5 that the discrepancy between results obtained from the numerical simulation and those obtained from the classical formulae is significant in the surrounding of the symmetry axis of the specimen. However, the estimation of errors following from the application of this assumption will be presented in Sect. 3.2.2. It seems that the formula (3.1) better approximates data from the numerical simulation than the investigated classical formulae. However, all distributions of the curvature radius of the longitudinal stress trajectory in the axis of symmetry tend to infinity. However, for the point laying in the plane of the minimum section of the sample on the free surface (so when $r = a$), the distribution takes the value of R. It is also worth noticing that in the scope of the classical approaches, the distributions of the curvature radii of the longitudinal

Fig. 3.5 Distribution of the curvature radius of the longitudinal stress trajectory in the minimal section plane of the axisymmetric sample for the chosen stage of deformation of ideal plastic (**a**) linear hardening (**b**) and nonlinear hardening (**c**) materials

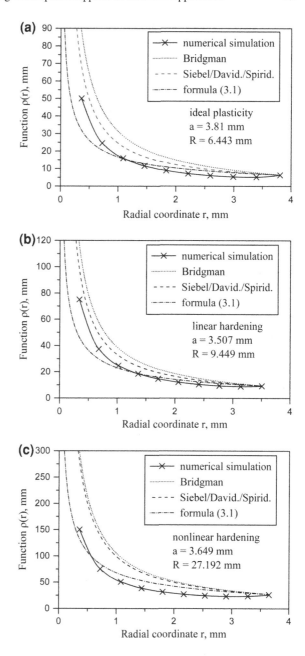

stress trajectories in the minimum section plane are strictly decreasing functions. Whereas, the radii ρ determined from the numerical simulation do not reveal monotonicity in a close surrounding of the neck contour. It is probably a result of the considerable strain location in this regime, which can destroy the monotonical behaviour of the distribution $\rho = \rho(r)$. This phenomenon is only slightly visible in the scope of Fig. 3.5 but it seems to be significant to underline.

3.2.2 Verification of Simplifications Due to Analytical Analysis

3.2.2.1 Assumption of the Equality of the Radial and Circumferential Stresses

The assumption of the radial and circumferential stress equality will be considered at the beginning assuming that the two remaining assumptions, i. e. whereas on the constancy of the yield stress in the minimum section plane of the sample and the form of the formula for the longitudinal stress trajectory, will be taken as valid. Because the circumferential and radial stresses do not need to be equal in the minimum section plane of the sample ($z = 0$), the relation between them can be written in a general form as:

$$\sigma_\theta(r) = \sigma_r(r) + \kappa c(r)\sigma_z(r) \quad \text{for} \quad 0 < r < a, \tag{3.2}$$

where κ should be a small parameter ($|\kappa| << 1$), and the unknown function $c(r)$ is normalised ($\max|c(r)| = 1$). It is obvious that before the neck creation, the equation $\sigma_\theta(r) = \sigma_r(r)$ is true and this means that κ is equal to zero. Therefore, the parameter κ can be considered as a measure of the plastic strain accumulation connected with the neck formation.

Taking advantage of the relation (3.2), the yield conditions according to Huber-Mises and Tresca, can be written after transformations in the form:

$$\sigma_z - \sigma_r = k + \upsilon\kappa c\sigma_z + O(\kappa^2) \quad \text{for} \quad 0 < \kappa \ll 1, \tag{3.3}$$

where the yield stress k is considered as constant with respect to r and the parameter υ has the value:

$$\upsilon_{HM} = \frac{1}{2}, \ \upsilon_T = 1, \tag{3.4}$$

respectively, for the conditions by Huber-Mises and Tresca. Differentiating the yield condition (3.3) in regard to r and introducing the results into the equilibrium condition, one can obtain the following differential equation in the minimum section plane:

$$(1 - \upsilon\kappa c)\frac{\partial\sigma_z}{\partial r} - \left(\upsilon\kappa c' + \frac{\kappa c}{r}\right)\sigma_z + \frac{k}{\rho} + \frac{1}{r}O(\kappa^2) = 0 \quad \text{for } 0 < \kappa \ll 1. \tag{3.5}$$

The solution of Eq. 3.5 together with the boundary condition on the free surface of the sample ($\sigma_r = 0$) is determined by the perturbations method with respect to the small parameter κ. Denoting the normalised axial stress occurring in the classical formulae by $\tilde{\sigma}_z^{(0)}$, the solution of Eq. 3.5 takes the form:

Table 3.1 Relative error arising from the introduction of the parameter κ for three chosen stages of deformation of ideal plastic material

Relative error (%)	Ideal plasticity		
	Stage of deformation a/R (–)		
	8E−05	0.185	1.509
Bridgman	−0.632	−6.657	−5.637
Siebel/David./Spirid.	−0.632	−6.656	−5.789
Formula (3.16)	−0.632	−11.018	−2.184

$$\frac{\sigma_z}{k} = \tilde{\sigma}_z^{(0)} + \upsilon \kappa c(r)\tilde{\sigma}_z^{(0)}(r) + \kappa \int_0^r \frac{c}{\xi}\tilde{\sigma}_z^{(0)}\,d\xi, \tag{3.6}$$

with an accuracy of order $O(\kappa^2)$. As a result, the difference between the classical formulae for the average axial stress and the above derived Eq. 3.6 equals:

$$\frac{\Delta\bar{\sigma}_z}{k} = \frac{2\kappa}{a^2}\int_0^a \upsilon c(t)\tilde{\sigma}_z^{(0)}(t)t\,dt + \frac{2\kappa}{a^2}\left(\frac{t^2}{2}\int_0^t \frac{c(r)}{r}\tilde{\sigma}_z^{(0)}(r)\,dr\Bigg|_0^a - \int_0^a \frac{r}{2}c(r)\tilde{\sigma}_z^{(0)}(r)\,dr\right). \tag{3.7}$$

Knowing the value of the small parameter κ, as well as the function $c(r)$ in Eq. 3.7, the errors of the classical formulae and of formula (3.16) connected with this assumption can be estimated. Therefore, the respective values κ and $c(r)$ were determined from the numerical simulation for the three different materials. It was observed that the value of parameter κ grows for small stages of deformation and takes its maximum value for parameter δ close to 0.2 and then decreases its value. On the other hand, the function $c(r)$ is increasing with respect to r for almost all stages of deformation and it changes its behaviour only for large plastic stains [1].

Integrating Eq. 3.7 under utilisation of the value range of parameter κ as well as function $c(r)$ obtained from the numerical simulation, the relative error of the average normalised axial stress can be determined. The highest error was obtained for the ideal plastic material. The appropriate data is shown in Table 3.1, from which it follows that there is practically no difference between the classical formulae. However, in the case of formula (3.16) for the average considered stage of deformation, the error is almost twice higher than for the classical formulae. Therefore, the assumption of the equality of the circumferential and radial stresses cannot be recognised as satisfied.

3.2.2.2 Assumption of the Constancy of the Yield Stress in the Minimal Section Plane

In the following, the simplifying assumption of the constancy of the yield stress in the minimum section plane will be considered, whereas the remaining two assumptions, namely the equality of the circumferential and radial stresses and the form of the formula for the curvature radius of the longitudinal stress trajectory, are taken as valid. Expanding in Taylor series, the function of the yield stress reads:

$$k(\varepsilon_{int}) = k(\bar{\varepsilon}_{int}) + k'(\bar{\varepsilon}_{int})(\varepsilon_{int} - \bar{\varepsilon}_{int}) + O(\varepsilon_{int} - \bar{\varepsilon}_{int})^2. \qquad (3.8)$$

Utilising the equilibrium equation 2.12 together with the boundary condition on the neck contour (2.75), we obtain:

$$\sigma_r = \int_r^a \frac{k(\varepsilon_{int})}{\rho(t)} \, dt, \qquad (3.9)$$

where $\varepsilon_{int} = \varepsilon_{int}(r)$ and $\bar{\varepsilon}_{int}$ is its average value in the minimal section plane. Inserting formula (3.8) into (3.9) and taking into account the yield stress

$$\sigma_z - \sigma_r = k(\varepsilon_{int}), \qquad (3.10)$$

written for both yield conditions in the same form (2.49), we obtain:

$$\sigma_z - k(\varepsilon_{int}) = \bar{k} \int_r^a \frac{dt}{\rho(t)} + \bar{\varepsilon}_{int} k'(\bar{\varepsilon}_{int}) \int_r^a \frac{\Delta \varepsilon_{int}(t)}{\rho(t)} \, dt + O((\Delta \varepsilon_{int})^2). \qquad (3.11)$$

In this equation the following expressions $\Delta \varepsilon_{int} = (\varepsilon_{int} - \bar{\varepsilon}_{int})/\bar{\varepsilon}_{int}$ and $\bar{k} = k(\bar{\varepsilon}_{int})$ were introduced. After a few transformations, the increase of the average normalised axial stress in comparison with the classical formulae and formula 3.16 is finally obtained in the form:

$$\frac{\Delta \bar{\sigma}_z}{\bar{k}} = \frac{1}{\bar{k}} k'(\bar{\varepsilon}_{int}) \left[\bar{\varepsilon}_{int} \frac{2}{a^2} \int_0^a r \left(\Delta \varepsilon_{int}(r) + \int_r^a \frac{\Delta \varepsilon_{int}(t)}{\rho(t)} \, dt \right) dr \right]. \qquad (3.12)$$

It should be noticed that the multiplicand on the right hand side of Eq. 3.12 is limited by one for the majority of flow curves utilised in plasticity theory and in engineering practice. This expression takes its largest value for the linear hardening material. Utilising the respective value of $\Delta \varepsilon_{int}$ obtained from the numerical simulation, the maximum relative error for the same value of parameter δ was determined [1]. The appropriate data is collected in Table 3.2. Again, there was no difference between the classical formulae and Eq. 3.16 observed. What is more interesting is that the error has a very small value not exceeding 0.5%. In addition, the value only slightly depends on the stage of plastic deformation. It follows that the assumption of the yield stress constancy in the minimal section plane can be considered as justified.

At this stage, it is worth noticing that the ideal plasticity case is characterised by a constant yield stress, therefore $k'(\varepsilon_{int}) \equiv 0$ and this is the reason this case was not considered in this analysis.

Table 3.2 Relative error arising from the yield stress variation in the minimal section plane of the sample for three chosen stages of deformation of the linear hardening material

Relative error (%)	Linear hardening		
	Stage of deformation a/R (–)		
	0.371	0.661	1.062
Bridgman	–0.358	–0.493	–0.481
Siebel/David./Spirid.	–0.358	–0.491	–0.476
Formula (3.16)	–0.348	–0.475	–0.455

3.2.2.3 Assumption of the Formula Form for the Curvature Radius of the Longitudinal Stress Trajectory

It is assumed in this consideration that both simplifications, i.e. equality of the circumferential and radial stresses as well as the constancy of the yield stress in the minimal section plane of sample, are valid, but the expressions for the curvature radius ρ (applied by Bridgman or Davidenkov-Spiridonova and in the approach presented in Sect. 2.8) are under question. It is worth noticing here that the distribution of the curvature radius of the longitudinal stress trajectory changes not only with the radial coordinate r, but generally speaking also with the stage of plastic deformation. As a result, it can be written that $\rho = \rho(r, \bar{\varepsilon}_{int})$. Therefore, the final formulae for the axial stress take the form [2]:

$$\sigma_z = k\left(1 + \int_r^a \frac{dt}{\rho(t, \bar{\varepsilon}_{int})}\right), \quad \frac{\bar{\sigma}_z}{k} = 1 + \frac{1}{a^2}\int_0^a \frac{r^2 dr}{\rho(r, \bar{\varepsilon}_{int})}, \tag{3.13}$$

and depend on the additional parameter $\bar{\varepsilon}_{int}$, which causes further difficulties in the verification process. Namely, it is essential to determine not only the radius distribution but also to deliver the results at each stage of the calculation. The curvature radius can be written in a general form (2.176) as:

$$\rho(r, \bar{\varepsilon}_{int}) = Ra\left(rG'(r^2/a^2)\right)^{-1}, \tag{3.14}$$

where the properties of function $G(t)$ were mentioned in Sect. 2.8. Generally speaking, this function is different at the each stage of deformation, i.e. $G(t) = G(t, \bar{\varepsilon}_{int})$, and the expression in the form of Eq. 2.203 can be used for the curvature approximation. Utilising the above assumption, the formula $(3.13)_2$ takes the form of Eq. 2.208.

Obviously, parameters $\alpha = \alpha(\bar{\varepsilon}_{int})$, $\beta = \beta(\bar{\varepsilon}_{int})$ can take different values for each calculation step. The expression (2.208) enables to estimate the error of the classical formulae and expression (3.16), connected with the choice of the curvature radius expression. The applied procedure consisted in calculating at each step the value of the curvature radius based on the numerical simulation utilising 10 nodes inside the range $(0, a)$. To this end, the definition of the radius based on the stress components was used under consideration of its numerical distribution in

Table 3.3 Relative errors arising from introduction $\rho(r)$ determined from the numerical simulation into the classical formulae and the formula (3.16) for three chosen stages of deformation of each material

Relative error (%)	Linear hardening Stage of deformation a/R (–)			Nonlinear hardening Stage of deformation a/R (–)			Ideal plasticity Stage of deformation a/R (–)		
	0.371	0.661	1.062	0.134	0.405	0.714	0.591	0.866	1.174
Bridgman	0.782	1.458	2.739	0.382	0.958	1.825	1.686	2.499	3.532
Siebel/David./Spirid.	0.302	0.112	–0.227	0.312	0.397	0.295	0.583	0.383	0.064
Formula (3.16)	–0.302	–0.900	–1.728	0.081	–0.257	–0.783	–0.332	–0.883	–1.555

Table 3.4 Maximum errors following from the considered simplifying assumptions

Simplifying assumption	Maximal error (%)	Error behaviour	Material	Stage of deformation a/R (–)
$\sigma_\theta = \sigma_r$	–11.018	↘	Ideal plasticity	0.185
$k = \text{const}$	–0.493	↘	Linear hardening	0.661
The form of formula $\rho(r)$	3.532	↗	Ideal plasticity	1.174

the surrounding of the minimum section plane. Subsequently, the curve was approximated by expression (3.14) together with (2.203) under consideration of the approximation accuracy. As a result, the values of these two auxiliary parameters, i.e. $\alpha = \alpha(\bar{\varepsilon}_{int})$, $\beta = \beta(\bar{\varepsilon}_{int})$, were determined at each calculating step [2]. Finally, the error was calculated (Table 3.3) between the classical formulae, formula (3.16) and the expression (2.208) for the chosen stages of plastic deformation.

It follows from the data presented in Table 3.3 that the relative error in the case of the Bridgman formula increases with increasing strain. The highest error occurs in the case of the ideal plastic material and is approximately 4% for the last stage of deformation. For the Siebel-Davidenkov-Spiridonova formula, the error has the same order but is lower than 0.6%. However, the relative error for formula (3.16) increases analogous to the Bridgman formula together with the increase of the deformation stage of the materials. Its highest value is reached for the last considered stage of deformation of the linear hardening material and is lower than 1.8%.

In order to easily compare all the obtained results in this subsection, Table 3.4 summarises the maximum errors determined during the verification of each simplifying assumption. Additionally, this table contains information on the type of material and the stage of deformation, at which these maximal errors occurred. The error distribution as a function of the stage of deformation is marked by arrows. Furthermore, the table contains the sign of the error because no normalisation was applied and only the relative errors were calculated between the formulae.

To sum up the above considerations, it should be underlined that each assumption influences the result in a different manner. The highest negative influence on the final result has the assumption of the radial and circumferential stresses equality and the formula form for the curvature radius of the longitudinal stress trajectory. On the other hand, the most justified assumption is the constancy of the yield stress in the minimum section plane.

3.3 Verification of the Formula for Determination of the Strain Intensity

The simplifying assumptions applied during the formulae derivation for the average normalised axial stress were verified in the previous subsections. However, the formula used among other by Bridgman for the determination of the invariant variable of the flow curve from the moment of the neck creation will be checked in this subsection. Instead of the strain intensity, he utilised the formula for the true (logarithmic) plastic strain (1.13).

It was mentioned in the Sect. 1.2 that the strain intensity is equal to the axial strain when one takes into account the material incompressibility condition and applies the simplification of the equality of the radial and circumferential strains. The assumption of the material incompressibility at great deformation stages does not raise any doubt, but, as it was mentioned, the simplifying assumption of equality of the radial and circumferential strains is questionable. It was calculated in Sect. 3.2 that the application of this simplification for the stresses determination can create even over 11% of error in the final result.

We intend to verify in this subsection the formula for the true plastic strain (1.13) utilising the strain intensity obtained from numerical simulation. At the beginning, the distributions of strain intensity in the minimum section plane of the specimen of each material were obtained from the numerical simulation. Then, their average values and the values following from the formula for the true plastic strain (1.13) were estimated. The appropriate data are placed in Fig. 3.6.

The conclusion can be drawn from Fig. 3.6 that results obtained from the formula for the true plastic strain (1.13) do not considerably differ from the average values of the strain intensity gained from the numerical simulation. In order to make the accuracy of approximation better visible, a diagram was prepared with the relative errors following from the approximation of the average strain intensity by the formula for the true plastic strain. The appropriate data conclude Fig. 3.7.

It follows from Fig. 3.7 that the application of the formula for the true plastic strain (1.13) for the determination of the average strain intensity generates only minor errors. Their highest values occur for smaller strains where the influence of the elastic strains is still large and the error maximally equals a value of –0.7% for the ideal plastic material.

Fig. 3.6 Comparison of the strain intensity obtained from the numerical simulation with results obtained from the formula for true plastic strain according to Eq. 1.13

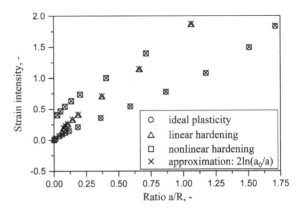

Fig. 3.7 Relative errors connected with the approximation of the average strain intensity obtained from the numerical simulation by the formula for true plastic strain according to Eq. 1.13

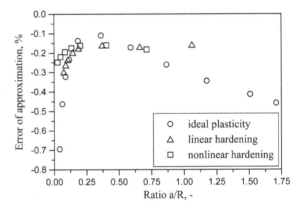

To sum up the considerations concluded in this subsection, it should be stated that the application of the formula for the true plastic strain as an invariant variable, instead of the strain intensity, in the determination of the flow curve is most of all justified.

3.4 Formulae Verification for the Average Normalised Axial Stress in the Minimal Section Plane

In this subsection, all the classical formulae and these derived in this monograph for the average normalised axial stress in the minimum section plane will be verified. Let us recall that in Sect. 2.8 a new formula was derived by generalisation of the classical formulae for the curvature radius of the longitudinal stress trajectory. This expression is given in Eq. 2.208 together with limitations for the parameters α and β, which values could not be determined analytically. For that

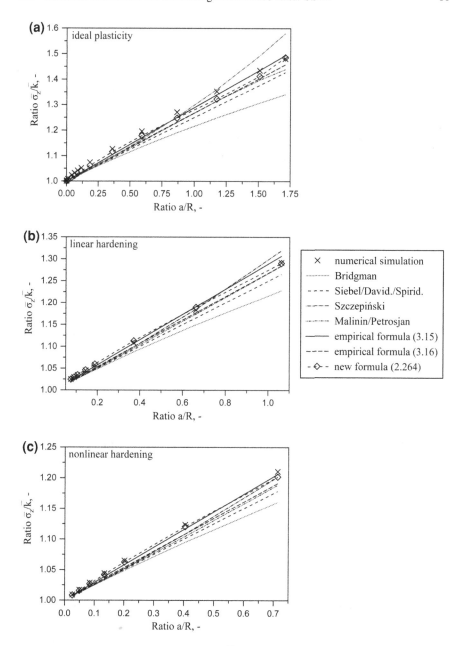

Fig. 3.8 Average normalised axial stress $\bar{\sigma}_z/\bar{k}$ as a function of deformation a/R for results obtained from the numerical simulation as well as the classical formulae, empirical ones (3.15) and (3.16) and also formula (2.264) for ideal plastic (**a**) linear hardening (**b**) and nonlinear hardening (**c**) materials

sake, this formula can be treated as an empirical expression and the values of parameters α and β can be estimated utilising the data obtained from the numerical simulation. Two sets of parameters α and β were suggested, where in the first case for $\alpha = 0.95$ and $\beta = 0.5$ formula (2.208) takes the form:

$$\frac{\bar{\sigma}_z}{k} = 1 + \frac{a}{4R} + \frac{19a}{488R}, \tag{3.15}$$

and revealed the best approximation of the average normalised axial stress got for all materials assigned in the numerical simulation. However, at small stages of material deformation the expression is a lower limit while for large strains it is an upper limit for the results. In the second case, the parameters were chosen as $\alpha = 0.5$ and $\beta = 0.5$ and formula (2.208) takes the form:

$$\frac{\bar{\sigma}_z}{k} = 1 + \frac{a}{4R} + \frac{a}{56R}, \tag{3.16}$$

and the expression becomes always a lower limit for the data obtained from the numerical simulation (Fig. 3.8).

Taking advantage of the relations for the average normalised axial stress from the classical approaches (2.81), (2.107), (2.133), and (2.172) and the empirical formulae (3.15) and (3.16) as well as the newly derived formula (2.264), relation $\bar{\sigma}_z/\bar{k}$ is presented in Fig. 3.8 as a function of the stage of deformation a/R for the three materials assigned in the numerical simulation. In addition to data obtained from the numerical simulation indicated by markers curves determined from the classical formulae and two empirical ones (3.15) and (3.16) (the solid line) were drawn. However, in the case of the newly derived formula (2.264), results directly obtained from this formula are indicated by markers. The dashed line was created as a result of the approximation of the relation $a_0/a = f(a/R)$ and its application in formula (2.264).

It is clearly visible from Fig. 3.8 that the new formula for the average normalised axial stress (2.264) better approximates data obtained from the numerical simulation for all materials if compared to the classical formulae.

The next step of the formulae verification for the average normalised axial stress is the reconstruction of the flow curves. The verification of the formula for the strain intensity determination demonstrated (cf. Sect. 3.3) that this simple formula for the true plastic strain accurately approximates results obtained from the numerical simulation. For that sake, its application during the flow curve determination is most of all justified. Whereas the yield stress is obtained by transformation of the formulae for the average normalised axial stress $\bar{\sigma}_z/\bar{k}$. Let us recall from Chap. 2 the form of all the mentioned classical formulae. For Bridgman, the formula for the yield stress has the form:

$$\bar{k} = \frac{\bar{\sigma}_z}{\left(1 + \frac{2R}{a}\right)\ln\left(1 + \frac{a}{2R}\right)}. \tag{3.17}$$

In the case of Siebel-Davidenkov-Spiridonova, the following form was obtained:

$$\bar{k} = \frac{\bar{\sigma}_z}{1 + \frac{a}{4R}}, \tag{3.18}$$

while for Szczepiński:

$$\bar{k} = \frac{\bar{\sigma}_z}{\frac{2R}{a}\left[\exp\left(\frac{a}{2R}\right) - 1\right]}, \tag{3.19}$$

and the approximated formula by Malinin-Petrosjan:

$$\bar{k} = \frac{\bar{\sigma}_z\left(1 - \frac{a}{8R} + \frac{5}{384}\left(\frac{a}{R}\right)^2 + \eta_1\left(\frac{a}{R}\right)^3 + \xi_0^{(0)}\eta_2\left(\frac{a}{R}\right)^4\right)^2}{\left(1 - \sqrt{2}\eta_2\left(\frac{a}{R}\right)^{\frac{9}{2}}\right)^2}, \tag{3.20}$$

with parameters $\xi_0^{(0)} \approx 2.404825558$, $\eta_1 = -0.001684$ and $\eta_2 = 0.00052196$.

Let us also present the form of two empirical formulae:

$$\bar{k} = \frac{\bar{\sigma}_z}{1 + \frac{a}{4R} + \frac{19a}{488R}}, \tag{3.21}$$

and:

$$\bar{k} = \frac{\bar{\sigma}_z}{1 + \frac{a}{4R} + \frac{a}{56R}}, \tag{3.22}$$

as well as the new formula:

$$\bar{k} = \bar{\sigma}_z\left[\begin{array}{l} 1 - \frac{5\Lambda}{7(1+5\Lambda)} - \frac{2(1-6\Lambda)}{7(1+5\Lambda)} + \frac{2}{7} + \frac{30\Lambda(8\Lambda - \delta - 5\delta\Lambda)}{49\delta(1+5\Lambda)^2} \\[3mm] + \frac{3(8\Lambda - \delta - 5\delta\Lambda)}{7\delta(1+5\Lambda)}\left(\frac{2(1-6\Lambda)}{7(1+5\Lambda)} - \frac{2}{7} - \frac{30\Lambda(8\Lambda - \delta - 5\delta\Lambda)}{49\delta(1+5\Lambda)^2}\right) \\[3mm] \ln\left|1 + \frac{7\delta(1+5\Lambda)}{3(8\Lambda - \delta - 5\delta\Lambda)}\right| \end{array}\right]^{-1}. \tag{3.23}$$

In addition to parameters a_0, a and R, the average axial stress in the minimum section plane $\bar{\sigma}_z$ corresponding to the true stress σ_{true} introduced in Sect. 1.2 occurs in expressions (3.17)–(3.23). The value of this stress can be calculated analogous to Eq. 1.4 as the ratio of the tensile force F to the surface area of the minimal section of the tension sample $S = \pi a^2$, $\bar{\sigma}_z = F/\pi a^2$.

Since we gained from the numerical simulation for each stage of deformation the distribution of the axial stress in the minimal section plane of the sample, their average values were determined and compared. It turned out that the highest error of the formula $\bar{\sigma}_z = F/\pi a^2$ application in comparison to the average values

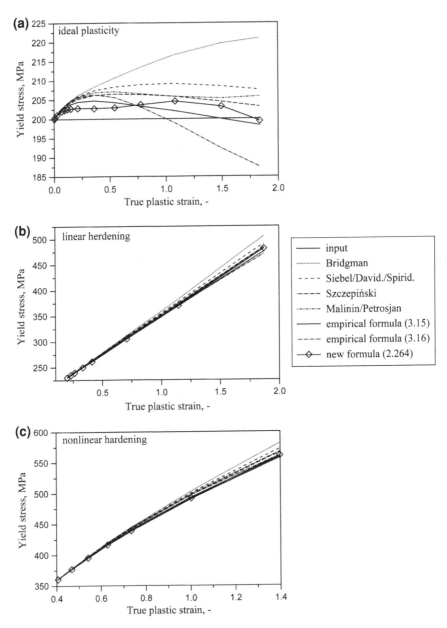

Fig. 3.9 Reconstructed flow curves for ideal plastic (**a**) linear hardening (**b**) nonlinear hardening (**c**) materials

estimated from the numerical simulation is equal to 0.39% in the case of the ideal plastic material. However, the error can be neglected for the linear and nonlinear hardening materials ($\sim 10^{-3}$%).

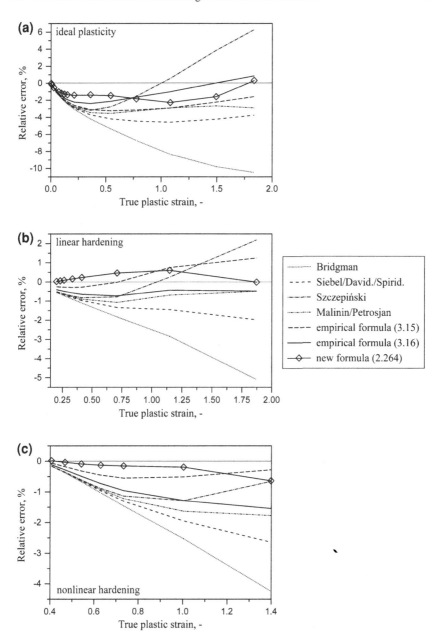

Fig. 3.10 Relative errors following from the application of the classical formulae, the empirical ones (3.15) and (3.16) and the formula (2.264) for the reconstruction of the flow curve assigned in the numerical simulation for the ideal plastic (**a**) linear hardening (**b**) and nonlinear hardening (**c**) materials

Table 3.5 Relative errors following from application of every formula for flow curve determination for the last considered stage of deformation

Relative error (%)	Linear hardening	Nonlinear hardening	Ideal plasticity
Bridgman	−5.1	−4.2	−10.5
Siebel/David./Spirid.	−2.0	−2.6	−3.8
Szczepiński	2.2	−0.7	6.3
Malinin/Petrosjan	−0.5	−1.7	−2.9
Empirical formula (3.15)	1.2	−0.3	0.9
Empirical formula (3.16)	−0.5	−1.5	−1.6
New formula (2.264)	0	−0.6	0.3

In addition to formulae (3.17)–(3.23), the average axial stress in the minimum section plane in the form $\bar{\sigma}_z = F/\pi a^2$ and the relation for the true plastic strain according to Eq. 1.13 was utilised for the determination of the flow curve. The obtained results for the flow curves are presented in Fig. 3.9.

The data included in Fig. 3.9 confirms that the yield stress obtained from the new formula is much more compatible with the assumed flow curves in the numerical simulation. Additionally, the absolute error following from the application of a particular formula for the reconstruction of the flow curve can be extracted. It follows from these drawings that the highest errors occurred for the ideal plastic material. In the case of the linear as well nonlinear hardening materials, the errors are lower. However, this illusion is caused by the scale of these diagrams. In order to make the accuracy of the reconstructed flow curves more visible, the diagrams of the relative errors are presented in Fig. 3.10.

It follows from Fig. 3.10 that the reconstruction accuracy of the flow curve by formula (2.264) is the same for all the considered materials. The errors following from the application of the formulae for the flow curve determination for the last considered stage of deformation are summarised in Table 3.5.

Analysing the results contained in Table 3.5, one can unequivocally confirm the greater approximation accuracy of the data obtained from the numerical simulation by the newly derived formula in comparison with the classical expressions.

References

1. M. Gromada, *Modelling of the plastic strain location in the tensile test (in Polish)* (Rzeszow University of Technology, Rzeszow, 2006)
2. M. Gromada, G. Mishuris, A. Öchsner, Necking in the tensile test. Correction formulae and related error estimation. Arch Metall Mater **52**, 231–238 (2007)
3. Program MSC.Marc Volume E: Demonstration Problems, Part II, Chapter 3 Plasticity and Creep
4. Standard EN 10002 –1 Metallic materials – Tensile testing – Part 1: Method of test at ambient temperature

Summary

On the base of the above considerations, one can draw the following most important conclusions regarding utilisation of formulae for properties reconstruction of elasto-plastic materials:

1. The formula derived by Bridgman for the flow curve determination demonstrates the worst results among all the classical formulae but it is still the most often used and disseminated especially in the English literature.
2. The formulae by Malinin–Petrosjan, Siebel–Davidenkov–Spiridonova and Szczepiński always characterise a greater accuracy compared to Bridgman. However, the smallest error among all the classical formulae generates the Malinin–Petrosjan formula. At small strains, the Szczepiński formula can be applied but at greater strains the Siebel–Davidenkov–Spiridonova should be used, because the curve determined from the Szczepiński formula begins to differ more from the real solution. However, the Siebel–Davidenkov– Spiridonova formula is still better in comparison with the results obtained from the Bridgman formula.
3. However, the most justified seems to be the application of the new formula which depends on two parameters. As it was presented, the new approximation always exhibits greater accuracy in comparison with the classical expressions.
4. If we are still interested in a formula taking into account only one parameter, it is worth taking advantage of the empirical formula (3.15) which reconstructs better the material properties. On the other hand, one may use the single-parameter formula (3.16) which reveals an acceptable accuracy but limits the real solution from below.
5. In the case of estimation of the flow curve of materials at small deformations and neck presence, it does not make sense to distinguish between the formulas because the difference for small strains is very small.

M. Gromada et al., *Correction Formulae for the Stress Distribution in Round Tensile Specimens at Neck Presence*, SpringerBriefs in Computational Mechanics, DOI: 10.1007/978-3-642-22134-7, © Magdalena Gromada 2011

Index

M. Gromada et al., *Correction Formulae for the Stress Distribution in Round Tensile*   89
Specimens at Neck Presence, SpringerBriefs in Computational Mechanics,
DOI: 10.1007/978-3-642-22134-7, © Magdalena Gromada 2011